Biogeography of the Southern End of the World

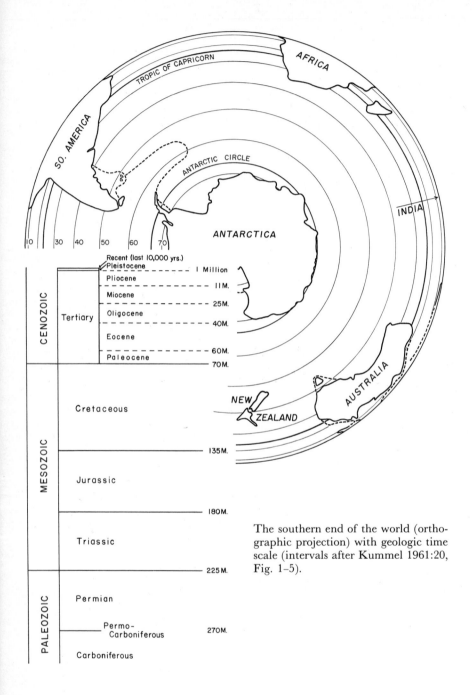

The southern end of the world (orthographic projection) with geologic time scale (intervals after Kummel 1961:20, Fig. 1–5).

Biogeography of the Southern End of the World

Distribution and history of far-southern life and land,
with an assessment of continental drift

PHILIP J. DARLINGTON, Jr.

Museum of Comparative Zoology, Harvard University

HARVARD UNIVERSITY PRESS

Cambridge, Massachusetts · 1965

Preface

This book is concerned primarily with the distribution, history, and significance of terrestrial plants and animals at the southern end of the world, in the southern cold-temperate zone and the antarctic region. However, the history of far-southern life depends on the history of the land. The two form one problem, and I shall try to solve it by evidence from nonbiologic sources as well as from plant and animal distribution.

The geographic distribution of plants and animals in the far south is increasingly exciting now. This is partly because knowledge of the Antarctic Continent and its history is increasing rapidly, partly because the theory of continental drift is finding new support from several sources, and partly because the distribution of far-southern plants and animals is exciting and instructive in itself. Widespread interest in the subject has resulted in much recent work, done in many different parts of the world and on many different groups of plants and animals, and much recent work has been done also on related subjects. If, three or four years ago, I had been able to order the particular pieces of work needed to prepare the way for a general southern biogeography, my order would have been largely filled! Among timely publications (listed in the Bibliography, pp. 219–229) are general papers on the distributions and histories of the floras and faunas of southern South America and New Zealand; important additions to knowledge of the distributions and histories of significant groups of southern plants and animals, including especially the southern beeches (*Nothofagus*); separate reviews of the geologic histories of South America, Tasmania, and New Zealand; additions to knowledge of the Antarctic Continent from many points of view resulting partly from the activities of the International Geophysical Year; and important general works on ancient climates, on the ocean bottoms (which may reveal the history of land), and on paleomagnetism.

My interest in southern biogeography began with my own field work (briefly described in Chapter 1) in Tasmania and southern Chile, but the present book brings together the results of much other recent work on many different subjects, some of them listed above. No subject can be treated exhaustively in the space available, but all are (I hope) adequately summarized and fitted into place in a coherent hypothesis of the history of the southern end of the world.

The importance of climate and of evolution in determining distributions of plants and animals will continually be stressed here, with new examples drawn mainly from predaceous ground beetles (Carabidae). If some of these insects are treated in more detail than might at first seem appropriate, this is deliberate and necessary. A point has been reached in southern biogeography from which advances can be made only by considering specific cases in detail. The cases are presented as examples of patterns of distribution and history that are common and significant. I should add that I still consider vertebrates the best animals for most biogeographic work but, for reasons that will be given, they fail in the far south, so that dependence must be placed on other animals and on plants.

Part I of this book describes existing situations and patterns of distribution of selected plants and animals in the far south. Part II treats worldwide patterns of evolution and dispersal and their relation to southern distribution patterns. This essay on the geography of evolution interrupts the discussion of southern biogeography but is necessary to an understanding of the southern patterns. Part III summarizes what is actually known of the geologic, climatic, and biotic histories of different far-southern lands. And Part IV is concerned more broadly with the history of life and land at the southern end of the world since the late Paleozoic, and with whether or not continental drift has occurred since then.

The style of a book like this should, I think, combine literal truth with simplicity. Every sentence should be calculated to convey a precise meaning that is exactly true, unambiguous, and expressed in the simplest appropriate words. I have tried to follow this rule. Editors have accused me of writing too simply. I have even been accused of using slang, which I do not do, but this accusation was blunted because the word that was objected to happened to be quoted from Darwin. Efforts to write the literal truth as simply as possible should have two results. One is, of course, to make the writing easily intelligible to readers. The other is to sharpen the understanding of the writer. I find that in trying to make every statement literally true and simple I discover gaps and errors in my own knowledge and deductions. In other words, in criticizing my own writing I also usefully criticize the ideas and conclusions I am writing about. For discussion of style I have often turned to *Plain Words: Their ABC* (by Sir Ernest Gowers, Knopf, New York, 1954), and for spellings, hyphenations, and abbreviations I have tried to follow *Webster's New International Dictionary* and *Webster's Geographical Dictionary,* and also the *Style Manual for Biological Journals* (American Institute of Biological Sciences, 2000 P St. N. W., Washington 6, D. C.).

In writing this book I have become indebted to more persons than I can list here. Some of them are named in appropriate places in the following pages. However, I must acknowledge special indebtedness to Miss Judith Hall for careful preliminary editing and final typing of the manuscript, to Miss Symme Burstein for painstaking work on the original maps and diagrams, and to my wife for help in many ways including proofreading. My work in Tasmania and Australia was supported by a John Simon Guggenheim Memorial Foundation fellowship, and that in southern South America, by National Science Foundation Grant GB-93.

The present small book follows (but does not repeat) my general *Zoogeography* (Wiley, New York, 1957), to which I shall refer continually in order to save space and avoid repetition. And my present conclusions have been summarized in a short paper in the *Proceedings of the National Academy of Sciences,* volume 52, pages 1084–1091 (October 1964).

P. J. D., Jr.

Contents

PART IV. INFERENCES ABOUT THE PAST

Biogeography of the Southern End of the World

1. Introduction

The different subjects brought together in this chapter have in common that they are all preliminary to my treatment of the biogeography of the southern end of the world. First is an initial statement of facts, principles, and procedures; then, a brief history of ideas about the distribution of plants and animals in the far south; then, a short account of my own work, especially in southern South America, Tasmania, and southeastern Australia; and finally, an introduction to beetles of the family Carabidae, because they supply so many of the examples used in later chapters.

Initial facts and principles. The initial facts and principles of distribution of plants and animals in the far south seem to me to be these. First, it is a fact that many plants and many invertebrate animals of the southern tip of South America (with Tierra del Fuego), the southern corner of Australia (with Tasmania), and New Zealand are related. Second, this pattern of geographic relationships within the southern cold-temperate zone occurs only among plants and invertebrate animals, not among terrestrial or fresh-water vertebrates (except some salt-tolerant fishes). Third, rainfall, cold, and other climatic and ecologic factors decisively affect distribution of life in the far south now and have probably done so in the past. Fourth, southern distributions should be considered in relation to the whole world. Fifth, it is important both to analyze single cases and to compare cases, to see whether different southern plants and animals have one common pattern of distribution, suggesting a common history, or whether they have diverse patterns, suggesting diverse histories. Each of these five facts or principles is worth further discussion. Several of them are subjects of later chapters in this book, but something more should be said here about land in the far south and about materials and methods in far-southern biogeography.

Land in the southern cold-temperate zone. The principal pieces of land in the southern cold-temperate zone (Frontispiece) are those named in the preceding paragraph: southern South America with Tierra del Fuego, southern Australia with Tasmania, and New Zealand. These lands have much in common geographically, climatically, and biologically, as will be shown, and no other substantial land areas share all their characteristics. The southern tip of Africa extends a little below the tropics but is warm-temperate rather than cold-temperate and differs in other ways (Chapter 11). And the Antarctic Continent, whatever it may have been like in the past (Chapter 12), is now ice-covered and virtually uninhabitable.

It should be remembered that Tierra del Fuego was connected to South America perhaps as recently as 10,000 years ago (Chapter 8) and that Tasmania was connected to Australia at about the same time (Chapter 9), during the last world-wide lowering of sea level that accompanied Pleistocene glaciation. Biologically, therefore, Tierra del Fuego is primarily an extension of southern South America rather than an independent archipelago, and Tasmania is an extension of southern Australia. In each case climate more than the very recent water gap has determined the nature and distribution of the southernmost flora and fauna. I do not know New Zealand and shall have comparatively little to say about it. The fact that it is a remote island which cannot draw freely on a continental biota must modify distribution patterns there. This fact suggests that comparisons should be made primarily between southern South America (plus Tierra del Fuego) and southern Australia (plus Tasmania) and that the comparisons should be extended to New Zealand cautiously, with due allowance for New Zealand's greater isolation.

Materials and methods in far-southern biogeography. No one volume can hold all that is now known about the distribution of plants and animals in the far south. A choice of materials and methods is necessary. One method is simply to collect as many cases as possible of relationships among plants and animals of southern South America, southern Australia, and New Zealand, express amazement at the total number of cases, and then make sweeping generalizations about southern land connections or continental drift. This is commonly done without stating premises or considering alternatives and without allowing for the profound effects of climate and evolution on the distribution of southern plants and animals. This method was perhaps the only one that could be used by Hooker more than a hundred years ago (but Hooker did consider alternatives) and the method was useful then, when few details were known and the general situation needed to be stressed. However, the method is out of date now.

Another method is to take one case or one class of cases, examine it exhaustively, and draw conclusions without reference to other cases. This is useful if conclusions are drawn with care and restraint. For example, it would probably be possible to fill a large book with what is now known about the distribution and history of the southern beeches, with pertinent information about their relationships, classification, reproduction, ecology, and means of dispersal, and with a careful statement of alternatives in attempting to trace their geographic history. Such a book would be of great value in southern biogeography. I hope that someone, perhaps Dr. Cranwell, will write it! However, the conclusions to be drawn from study of any single group are obviously limited.

A third method of procedure in far-southern biogeography is to select a set of appropriate cases that are adequately known and rather diverse, present them in sufficient detail but not exhaustively, and then compare them and see what the comparison shows that single cases do not show. This seems likely to be the most rewarding method, and it is the one I have chosen to follow here. The plants and animals selected for reasonably detailed consideration and comparison are the southern beeches (*Nothofagus*), phytophagous peloridiid bugs, and certain tribes and genera of predaceous carabid beetles. These particular plants and insects have been dealt with repeatedly by biogeographers in the past, but new and in part unexpected facts have been discovered about them in the last few years. Besides these groups, mammals and fresh-water fishes will be discussed (Chapter 7) for what they tell of the general history of South America and Australia, although they tell little directly about the cold southern tips of the continents, and other groups of plants and animals will be referred to more briefly in appropriate places.

History of southern biogeography. The distribution of plants and animals at the southern end of the world has interested biogeographers ever since biogeography has been a science. Hooker, in *Flora Antarctica* and related works published from 1844 to 1860 (see Turrill 1953), emphasized relationships among the floras of southern South America, Tasmania, and New Zealand. He had visited all these places. He thought that "land communications" between the areas in question were required for the higher orders of plants, but that spore-bearing forms might have been carried long distances by the "violent and prevailing westerly winds." And he thought that cold-adapted, subantarctic plants isolated on mountains on New Zealand must have dispersed while the climate was colder than now and must have been stranded on mountain tops by a change (warming) of climate. The land communications are very doubtful; the change of climate is certainly correct in general.

Darwin was the first thoroughgoing biogeographer (see Darlington 1959b). His voyage preceded Hooker's, although the quotation given below was published later than most of Hooker's work. Darwin mentioned far-southern distributions, but his explanation was different from Hooker's. He said in *The Origin of Species* (1859, 1964:381–382):

I am inclined to look in the southern, as in the northern hemisphere, to a former and warmer period, before the commencement of the Glacial period, when the antarctic lands, now covered with ice, supported a highly peculiar and isolated flora. I suspect that before this flora was exterminated by the Glacial epoch, a few forms were widely dispersed to various points of the southern hemisphere by occasional means of transport [across the sea], and by the aid, as halting-places, of existing and now sunken islands, and perhaps at the commencement of the Glacial period, by icebergs.

This explanation is still favored by many biogeographers. It was a remarkable deduction on Darwin's part, for he did not have the evidence we have now of fossil forests on Antarctica. This is an example not only of Darwin's prescience but also of the fruitfulness of enabling a good biologist to see things for himself. During the voyage of the *Beagle* Darwin saw for himself, as Hooker did, how plants and animals are distributed in the far south, and he realized that the geographic patterns he saw could not be accidental and had to be explained. To help young biologists go wherever they wish to go and see things for themselves is still a fruitful policy, of course. But this is a digression.

The history of southern biogeography since Hooker and Darwin can be outlined only very briefly here. Real advances have come primarily as a result of increase of detailed factual knowledge, which has been due to the efforts of a very large number of collectors and taxonomists working in different places and on different groups of plants and animals, living and fossil. This work has not been dramatic and the men and women who have done it are little known, but the importance of their work should be remembered. To write a history of southern biogeography entirely in terms of biogeographic ideas, without noting the great increase and diversification of factual knowledge, would be like writing a history of nations entirely in terms of personalities, without noting increase of populations or economic factors. We know enormously more now than Hooker and Darwin did about details of distribution in the far south, about probable relationships and phylogenies of southern plants and animals, and about distributions in the past. We also know enormously more about the geologic histories of southern lands (Chapters 8 to 13) and about the probable or possible relation of plant and animal distribution to climate and evolution (Chapters 5 and 6).

On a more theoretical level there has been a long-running, three-cornered argument in southern biogeography. Many persons, like Hooker, have thought that dispersal must have occurred across land connections in the far south. Others, like Darwin, have postulated dispersal across far-southern water gaps. And others, notably Matthew and Simpson, have emphasized that discontinuous southern distributions have often been formed by expansion and contraction of ranges, by groups of plants and animals first spreading over the world via northern routes, then dying back into discontinuous southern areas. To call the argument three-cornered is, of course, an oversimplification. Many persons, including, I think, all those just named, have realized that dispersal is complex and that different groups may have dispersed in different ways in the south as elsewhere.

Actually, southern biogeographers have been concerned chiefly with two questions. One question is whether specific groups of plants or animals, for example marsupials, have dispersed by northern or by southern routes. The other question is whether, if dispersal has occurred by southern routes (as nearly everyone agrees it has in some cases), it has been across former land connections or across water gaps. And believers in southern land connections have argued among themselves whether the hypothetical connections were narrow land bridges (isthmian links) or broad contacts between continents preceding continental drift.

Of the many papers that have contributed to or reviewed these arguments in the last hundred years, I need mention only two now. One, by Wittmann (1934), is concerned with the whole southern hemisphere and includes an extensive review of earlier publications on southern biogeography. The other, by Simpson (1940a), is concerned primarily with the extreme south and is less a review than an analysis of problems and a revaluation of ideas and evidence about the role of Antarctica in the past. I shall attempt no detailed review of more recent literature. Important recent works are noted in appropriate places in the following pages, and a very short history of ideas about continental drift is given in Chapter 18.

All the arguments mentioned above are still continuing. Nothing has been settled. That is why I undertook to write this book: to see whether, with all the new knowledge that has become available, the fundamental problems of far-southern biogeography cannot be solved or at least clarified.

My own work in the southern cold-temperate zone. Part of what I now have to say is a result of my own recent field work in extreme southern South America during seven weeks from November 25, 1962, to January 13, 1963. The time was spent wholly in Chile. Thanks to good advice from two capable Chilean entomologists (Prof. Guillermo Kuschel and Sr. Luís Peña) and to the great courtesy and aid of the Chilean Air Force and Navy, I was able to sample all the principal climatic and biotic regions of South America south of latitude 49° (Fig. 2): open steppe (east of Punta Arenas), deciduous beech forest (Rio Rubens and Puerto Williams), wet subantarctic moorland with patches of evergreen beech forest (Puerto Edén and Orange Bay), and the mountain zone (above Puerto Williams). The southernmost locality reached, Orange Bay, is about 50 km (about 30 mi) north of the latitude of Cape Horn.

Travel in the far south of South America provides both suspense and drama from time to time, mostly because of the weather. Suspense, for me, came during nine days that I waited at Punta Arenas while wind threatened the success of my work. It prevented me from flying across the main island of Tierra del

Fuego, and across the Beagle Channel, to Puerto Williams. Of course this was the same wind that used to keep sailing ships from rounding the Horn toward the west. It was sometimes quicker to sail eastward around the world at that latitude than to wait to pass Cape Horn! And drama, for me, came in landing at Orange Bay in the face of a driving snow storm on the second day of January, which is equivalent to the second day of July in the northern hemisphere. I mention these incidents because they exemplify the climate that plants and animals must tolerate at the southern tip of South America. An account of other "adventures" would be out of place here. Useful collections of Carabidae were made at all the localities named above, but the great value of my work was that it enabled me to see for myself how plants and animals are distributed in relation to latitude and climate in southernmost South America.

My recent work in southern South America complements more extensive work done in Tasmania and eastern Australia (Fig. 3) during 19 months in 1956–1958 (Darlington 1960b). During most of this time I traveled and lived in a small truck with my wife and son, who was 14 years old when we started. Five months were spent in Tasmania, where I saw, collected in, and often camped in all the important pieces of wet south-temperate rain forest within reach of a vehicle or within reasonable reach on foot. The southernmost locality visited was Cockle Creek, about 8 km (5 mi) from the actual southern tip of land. Beyond this, southward, there is no land short of the Antarctic Continent. On the mainland of eastern Australia I worked in every important piece of wet forest from the Otway region (southwest of Melbourne) east and north to the tip of Cape York, so that I was able to see where and how south-temperate forests and south-temperate groups of animals (especially Carabidae) occurred, how they decreased in numbers northward, and how they formed transition with tropical groups. This transition as it occurs among certain Carabidae is described elsewhere (Darlington 1961a), in a paper referred to again in the following pages.

In short, my field work enables me, of my own knowledge, to compare the general situations, forests, and some insects of southernmost South America (with Tierra del Fuego) and southernmost Australia (with Tasmania), two widely separated pieces of land that are in many ways extraordinarily similar.

Carabid beetles. Beetles of the family Carabidae supply a number of examples of distributions discussed in the following pages. These insects have been used in zoogeographic work before, in recent years especially by Jeannel, Lindroth, and Britton, as well as by me (see Bibliography), and they might well be used more extensively in the future. They are among the most useful zoogeographic indicators in situations where vertebrates fail, especially in small, isolated, inhospitable areas. For these reasons an introductory account of Carabidae is worth giving here at some length. I should add that I am personally familiar with them. I have studied their taxonomy for more than thirty years, and I have collected them extensively in Tasmania and Australia, southern South America, and some other parts of the world.

Carabidae, or "ground beetles" (Figs. 8 and 10) are typically rather unspecialized, active, ground-living, predaceous beetles. The family is characterized technically by division of the first ventral abdominal segment, legs adapted for running or digging (not swimming), wings (if fully formed) with characteristic venation, and many other details that need not be given now. Different carabids differ greatly in size, from less than 1 to at least 65 or 70 mm in length. Different carabids differ also in adaptations, ecologic requirements, and habits. They include surface-running, fossorial, and arboreal groups. Some live in forest and some in open places, and some on wet and some on dry ground, and of the wet-ground forms some live only beside running water while others live in various other wet places. Special, limited carabid faunas occur on mountain tops and isolated islands (Darlington 1943). Some carabids have entered caves and become blind and otherwise highly modified there, and others have invaded the intertidal zone along seacoasts. Some carabids can fly and some cannot, and this difference is often reflected in their distributions (see following pages). Different carabids differ also to some extent in food requirements, for a few are phytophagous and a few myrmecophilous, and the larvae of a few are parasitic on other insects. However, most Carabidae appear to be rather simple predators, preying both as larvae and as adults on insects and other invertebrates. Their distributions usually are not limited by dependence on single species of other animals or of plants.

About 30,000 species of Carabidae are known, more than of all vertebrates together excepting marine fishes, and hundreds of

additional species are discovered and described each year. Carabids are abundant on all continents except the Antarctic, and on
most islands, and in all climates from the tropics north to the
arctic and south to the subantarctic. Nevertheless different groups
within the family are very differently distributed, and their distributions are often correlated with climate.

That climate limits the distribution of many Carabidae is a
fundamental fact in their zoogeography. Many dominant groups
are wholly or mainly tropical. Some other groups, including some
large genera and even some tribes, are confined to the north
temperate zone, and their southern limits are often closely correlated with climate. The genus *Carabus* is an example. It is a
dominant genus, of hundreds of species, and the insects are relatively large, conspicuous, and therefore well known. Most are now
entirely flightless, but winged individuals occur in a few species
(see below), and the whole genus is evidently derived from winged
ancestors, which could disperse by flight. Nevertheless, although
it occurs around the world north of the tropics, and although
some species occur south to North Africa, south-central Asia including the Himalayas and northern Burma, northern Florida,
and northern Mexico (Fig. 1), *Carabus* apparently nowhere enters
the full tropics. Moreover no *Carabus* have been found stranded

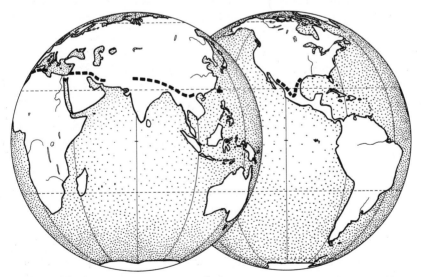

Fig. 1. Approximate southern limits of distribution of the dominant northern
ground-beetle genus *Carabus*.

on mountain tops deep in the tropics, which suggests that the genus has always been strictly nontropical. This is, as I have said, an example. A number of other groups of Carabidae and many other insects are distributed in more or less the same way, ranging around the northern part of the world but not entering the tropics. The existence of so many strictly nontropical groups in the Northern Hemisphere emphasizes the importance of the climatic zones in insect distribution and suggests that zonation of climate may be an important factor in determining distributions in the Southern Hemisphere too.

All Carabidae are descended from winged, flying ancestors, and the majority of existing species are still winged. However, the wings have atrophied and the insects therefore have become flightless in many different groups of these beetles in many different parts of the world. The powers of dispersal of different Carabidae vary with the insects' size and power of flight. Very small flying forms can get into the air under their own power but are then carried almost involuntarily by wind, and they may be blown great distances. Carabidae of this sort have apparently sometimes been carried by prevailing winds even across the Atlantic Ocean, from North Africa to the West Indies (Darlington 1938b). Very small Carabidae, like other very small arthropods, may be wind-dispersed even if they cannot fly. Some larger, winged Carabidae fly strongly and for long distances. On the other hand, large, flightless Carabidae have relatively low powers of dispersal and do not easily cross even narrow barriers. This difference is shown, for example, by the relationships of different Carabidae of New Guinea and of the adjacent northeastern corner of Australia. New Guinea and Australia were connected by land very recently (see Chapter 9), but movement of forest-living insects across the connection was apparently hindered by barriers of relatively open, relatively dry country. The result is that, although a number of forest-living Carabidae are identical in New Guinea and northeastern Australia, almost all of them are winged species that evidently flew across the ecologic barriers. At the same time, none of the many flightless Carabidae of Australia succeeded in reaching New Guinea, and the only flightless New Guinean species that reached Australia was an arboreal tiger beetle which (judging from its wide distribution in the Malay Archipelago) may have come by sea, on drifting trees, rather than by land (Darlington 1961a). The evidence is clear: in general, winged forest-living Carabidae did and flightless ones did not cross the ecologic barriers that separated New Guinea and Australia when a land connection existed.

Wings of Carabidae atrophy by mutation, and mutation results first in wing dimorphism, with long-winged flying individuals and short-winged flightless ones occurring in the same population. Dimorphically winged species of Carabidae are numerous and probably occur in every part of the world. Long and short wings are inherited in Mendelian fashion, with short wings dominant, at least in some cases (Lindroth 1946). The recessive allelomorph for long wings sometimes persists in short-winged populations for considerable times. Long-winged individuals may therefore appear unexpectedly in apparently short-winged, flightless species and genera, and such individuals may fly. This is the case, for example, in the genus *Carabus* (see above). Of the hundreds of existing species of this genus none is constantly long-winged. Every species of the genus is normally flightless. Nevertheless, long-winged individuals still appear in several European and North American species, and they do fly (Darlington 1936:154; Lindroth 1963a:31). This kind of dimorphism, with long-winged flying individuals appearing unexpectedly in supposedly flightless groups, occurs among some other insects, including the primitive southern Homopteran family Peloridiidae (pp. 31–35).

Dimorphically winged Carabidae are especially numerous in northern regions that were glaciated during the Pleistocene, including northern Europe and northeastern North America. Scandinavia and Finland together have at least 50 dimorphic species (Lindroth 1949:337–338). In many cases wing dimorphism seems to be correlated with instability of climate. Short-winged individuals apparently have had an advantage in small refugia during glaciations; long-winged ones, when ice has retreated and new ground has been colonized. The recent geographic histories of many northern species are clearly shown by the distributions of the wing forms: short-winged individuals predominate where a species survived the last glaciation, and long-winged individuals (which Lindroth calls "parachutists") predominate where the species has spread, by flight, into recently deglaciated country. (See Lindroth 1949:335–416; 1963b:96–102 for further details and for maps of the distribution of the wing forms of many dimorphic species in Scandinavia and Newfoundland.) Wing dimorphism occurs in some species of Carabidae in southern South America and Tasmania too. These species have not been adequately studied. They would probably repay study.

The geographic history of Carabidae has to be deduced mainly from indirect evidence, for their fossil record is very poor. Fossil carabids in stone are virtually useless. They rarely show the technical characters on which classification of the insects is based. A few Carabidae are fossil in the Baltic amber, but they have not been adequately studied. Remains of many northern Carabidae have been found in Pleistocene interglacial peat, especially in northern Europe. Most are fragmentary, but they are identifiable and some have been carefully studied by competent specialists. They show nothing directly about the history of southern Carabidae, but they do show that there has been little or no perceptible evolution of northern species since Pleistocene interglacial times (Lindroth, personal communication, 1963). Southern carabid species presumably evolve no more rapidly than northern ones. Therefore, very distinct species and genera of Carabidae now found localized in southern cold-temperate areas presumably date from the Tertiary or before.

In summary: although Carabidae occur almost everywhere, different ones are specialized in many different ways. They are usually not dependent on particular species of other animals or of plants, but their distributions do depend on ecologic factors. Climate, especially the difference between tropical and temperate climates, limits the distributions of many of them. Some fly and some do not, and their powers of dispersal vary accordingly. Many otherwise flightless groups of Carabidae, including some far-southern ones discussed in later chapters, still include exceptional winged or dimorphic species, suggesting that the groups dispersed by flight before wing atrophy occurred. Present flightlessness does not rule out dispersal by flight in the past. This is a special case of a general rule: the tolerances and powers of dispersal of existing plants and animals are not necessarily the same as the tolerances and powers of dispersal of their ancestors.

PART I   Existing geographic patterns

2. Distribution in relation to climate in the far south

The profound effect of climatic (ecologic) factors on distribution of plants and animals in the southern cold-temperate zone must be seen to be fully appreciated. Seeing it—that is, seeing plant and animal distributions at the southern tip of South America in relation to climate—was the most important result of my recent visit. Two different climatic factors have conspicuous effects on distribution in the far south. One is rainfall. The other is cold, or rather cold plus a set of related factors that make parts of southernmost South America relatively inhospitable to life.

Zonation with rainfall. The importance of rainfall is emphasized by a map of the major faunistic regions of southern Chile in an excellent short paper on the terrestrial zoology of the region recently published by Professor Kuschel (1960:542; my Fig. 2).

Most wind and rain come from the west in southern South America, and much of the rain falls on the western edge of the land. There is therefore a narrow, almost continuous zone of heavy, wet, south-temperate rain forest extending from north to south along the west coast from central Chile almost to Cape Horn, a distance of more than 2000 km (about 1300 mi). This forest includes southern beeches of the genus *Nothofagus,* and toward its southern end much of the forest consists of almost pure stands of single species of *Nothofagus.* East of the forest, where there is much less rain, is a zone or triangle of open steppe extending from the warm-temperate part of Argentina south to northeastern Tierra del Fuego. And on the other side of the forest, along the extreme western edge of land in the far south, is a third, narrow zone of open, very windy, very wet, and (in the farthest south) very cold subantarctic moorland. There is also, above the forest line in western South America, a mountain zone including mountain moorland.

The situation in Tasmania (Fig. 3) is generally similar to that

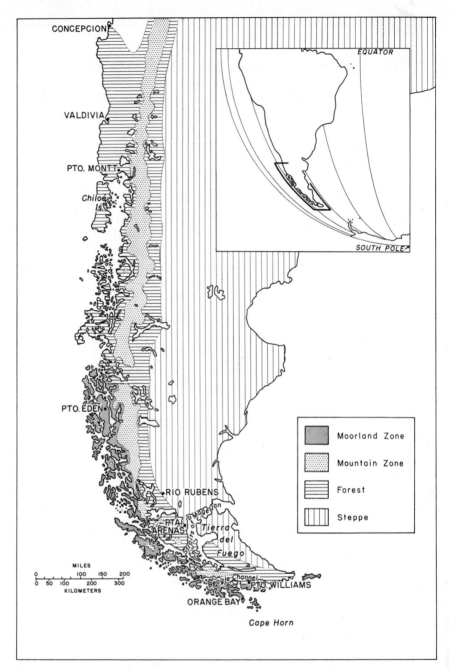

Fig. 2. Southernmost South America, showing zonation of vegetation determined by zonation of rainfall (from Kuschel 1960, Fig. 48, redrawn and simplified) and principal localities named in the text.

in southern South America. Wind and rain tend to come from the west, and much of western Tasmania is covered with heavy, wet, closed, south-temperate rain forest, including *Nothofagus,* while most of eastern Tasmania is covered with drier, open forest dominated by *Eucalyptus* and without *Nothofagus.* However, Tasmania does not lie as far south as southern South America. Rain

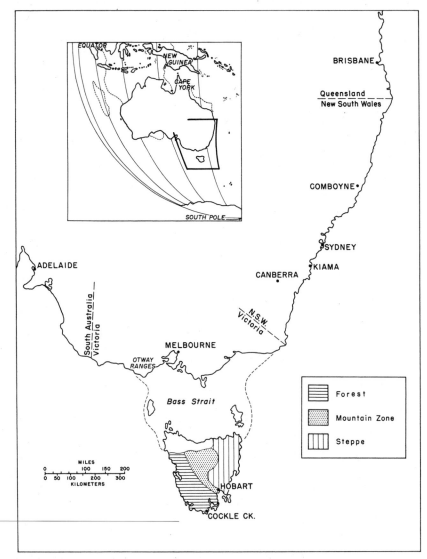

Fig. 3. Tasmania and southeastern Australia, showing zonation of vegetation on Tasmania (diagrammatic) and principal localities named in the text.

and vegetation zones are less sharply marked. And open steppe and low-altitude moorland are lacking or poorly developed, although mountain moors occur above the forest line.

The situation on New Zealand seems to be generally similar too, with most wind and rain coming from the west, and with the western side of South Island originally almost entirely covered with heavy forest consisting in many places (but not everywhere) of *Nothofagus*. However, some other parts of New Zealand are extensively forested too, or were before being cleared. Moorland occurs near sea level in southern New Zealand and on mountains farther north.

Many or perhaps most groups of plants and invertebrates that, like *Nothofagus*, occur in both southern South America and Tasmania are confined to the wet forest or moorland on the *western* side of each piece of land. The steppe of southeastern South America and the drier open forest of eastern Tasmania have very different floras and invertebrate faunas, derived mainly from the warmer parts of South America and Australia respectively. I have discussed this fact at greater length elsewhere (1960a:660–661, Fig. 80). This pattern of relationships emphasizes the importance of climate in distribution of far-southern plants and animals. The pattern is diagrammed in Fig. 4.

An example of a southern cold-temperate group of insects that, in southern South America and Tasmania, occurs only in the western areas of heavy rainfall is the tribe Migadopini of carabid beetles. Four genera of this tribe occur in extreme southern South America. Two of them are characteristic of the western forest zone and two of the subantartic moorland (see again Fig. 2), and there is no migadopine in the drier, opener steppe country in the far south. Two other genera occur in Tasmania, and both are, on the island, confined to the wet western forests, although their distributions differ in detail and one of them has an additional species isolated in the Otway Forest on the Australian mainland. There is no migadopine in the drier, opener forest of eastern Tasmania. This statement of the distribution of the Migadopini is true as far as it goes but is incomplete. The full distribution of the tribe is given in the following chapter.

Diminution of flora and fauna southward. Cold and related climatic factors that increase in intensity toward the south (presumably

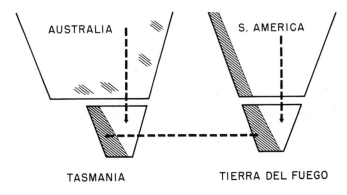

Fig. 4. Diagram of geographic relationships of many plants and invertebrates of wet forest (hatched areas) and of more open country (not hatched) in southern South America and Tasmania.

alternation of seasons, shortness of days in winter, and lack of warmth in summer, as well as direct cold) make a second fundamental pattern in the distribution of plants and animals in southern South America: decrease of diversity southward. The forest itself exemplifies this. The northern end of the Chilean forest is in a warm-temperate climate, and the forest there consists of many different kinds of trees, with many lianas and epiphytes. Southward, although the forest is virtually continuous, its diversity decreases until, south of the Straits of Magellan, only six species of trees remain (three of them species of *Nothofagus*), lianas are absent, and epiphytes are few (Kuschel 1960:543–544).

The fauna of the forest diminishes too. Many mammals, including three genera (two of them endemic) of marsupials, occur in the forest at its northern end, but no marsupials and few other mammals reach the southern end of the virtually continuous forest strip. In fact only the following terrestrial mammals are known to occur as far south as Tierra del Fuego in all habitats combined: one amphibious otter, one fox, one "camel" (the Guanaco), and one cavioid and several cricetid rodents (Osgood 1943:30–31); and see pages 66–67 of the present book for further details of the distribution of southernmost South American mammals in relation to habitat.

Amphibians and reptiles decrease southward even more strikingly. Many frogs, including endemic genera, inhabit the northern end of the Chilean forest, and I found four species of frogs in forest and on moorland at Puerto Edén, about 49° S. But in the

extreme south, for example at Puerto Williams and Orange Bay
(see again Fig. 2), amphibians are entirely absent both in the forest and on the moorland. The southernmost amphibians are actually one or two small frogs that reach the opener, warmer part of
northeastern Tierra del Fuego. One or two lizards reach this part
of Tierra del Fuego too, and this is the southern limit of reptiles.
Reptiles as well as amphibians are entirely absent in the southernmost forest and on the southernmost moorland. The diminution
of the reptile fauna southward is illustrated by the decrease of
numbers of snakes from north to south in Argentina (Table 1).

Fishes of strictly fresh-water families decrease in numbers southward even more rapidly than the amphibians and reptiles. The
Amazon River and its tributaries possess the richest fresh-water
fish fauna in the world. It is dominated by characins and catfishes
but includes other groups of fishes in smaller numbers. Toward
the south, however, subtractions from the Amazonian fish fauna
begin even in southern Brazil. Many tropical fishes follow the
Paraná–La Plata River system southward into northern Argentina,
but only a few characins and catfishes occur farther south. The
last characin may reach only 41° 18′ S. Stream-living pygidiid
catfishes reach at least 47° 30′ S, but they are the last, southernmost representatives of the tropical fauna of true fresh-water
fishes, and their final limit is about 900 km (about 600 mi) above
the latitude of Cape Horn. South of this, on the southern tip of
South America, there are no fishes of strictly fresh-water groups.
However, a few very different fishes, of salt-tolerant groups, do
occur in fresh water south even to Tierra del Fuego. Most of these

TABLE 1. Decrease of number of species of snakes southward in eastern Argentina (from Serié 1936).

Area	Approximate S latitude	Number of species
Misiones	26°–28°	55
Corrientes	28°–30°	51
Entre Rios	30°–33°	32
Buenos Aires	33°–39°	22
La Pampa	35°–40°	15
Rio Negro	39°–42°	5
Chubut	42°–46°	5
Santa Cruz	46°–52°	1
Tierra del Fuego	53°–55°	0

fishes belong to a few families that do not occur in the tropics but that are represented in southern Australia and New Zealand as well as southern South America (see pages 38, 82). The change from the very rich fish fauna in the tropics of South America to the very limited fauna, derived from different families, at the southern tip of the continent emphasizes the fundamental importance of climate in distribution even of animals in fresh water.

Even land birds are much reduced in numbers toward the southern tip of South America. The following South American families, for example, apparently do not reach Tierra del Fuego, and most of them do not even approach the southern tip of the continent very closely: rheas (south to the Straits of Magellan), tinamous, cracids, trogons, puff birds, jacamars, toucans, ant birds (of which more than 200 species are known in the warmer part of South and Central America), woodhewers, manakins, cotingas, and others. And of the following additional families that are well represented in tropical South America apparently only single species reach Tierra del Fuego: parrots, pigeons, hummingbirds, and woodpeckers. Crawshay (1907) lists 40 species of land birds from Tierra del Fuego: 7 hawks, 5 owls, 1 parrot, 1 woodpecker, and 26 passerines. This list is not quite complete—Crawshay missed the Fuegian hummingbird and a dove, for example—but it is enough to show that the land birds of Tierra del Fuego number only between 10 and 20 percent as many species as occur in reasonably diverse areas of roughly the same size in tropical America, for example in the tropical part of the Santa Marta region (Todd and Carriker 1922) or in the Panama Canal Zone (Sturgis 1928).

The insect fauna diminishes strikingly too, and the diminution occurs in some of the most dominant groups of South American insects. For example, 200 or more species of ants may be found at single localities in the Amazon rain forest, and ants are abundant elsewhere in the warmer parts of South America, but they diminish progressively southward (Table 2) until only two species are known south of the Beagle Channel, and apparently none reaches the southernmost moorland.

Carabid beetles diminish southward too. Of the specialized, mostly diurnal, conspicuous hunting carabids called tiger beetles (subfamily Cicindelinae), of which there is a large fauna in the warmer part of South America, only one species occurs as far south as the Straits of Magellan. The almost cosmopolitan tribe

TABLE 2. Decrease of number of species of ants southward in South America (from Kusnezov 1957).

Area	Approximate S latitude	Number of species
São Paulo, Brazil	20°–25°	222
Misiones, Argentina	26°–28°	191
Tucumán, Argentina	26°–28°	139
Buenos Aires, Argentina	33°–39°	103
Patagonia as a whole	39°–52°	59
Patagonia, humid west	40°–52°	19
Tierra del Fuego	53°–55°	2

Scaritini, which includes fossorial Carabidae of various sizes, and which is represented in the warmer parts of South America by about 24 genera and probably more than 200 known species, apparently does not reach the Straits at all. The cosmopolitan tribe Harpalini, which includes a diversity of small and medium-sized, ground-living species in warmer climates, reaches Tierra del Fuego, but only one or two species do so, and they seem to occur only in the opener, warmer habitats there, not in the forest or moorland. The cosmopolitan, primarily ground-living tribe Pterostichini, which is well represented elsewhere in South America, is poorly represented in the far south and reaches only the warmer, opener places there, not the southernmost forest or moorland. And *Tachys,* an almost cosmopolitan genus of small, damp-ground-living carabids which swarm in all warm regions including the warmer part of South America, apparently does not reach the Straits of Magellan. In the vicinity of Punta Arenas the place of *Tachys* is taken by *Bembidion,* which has an essentially bizonal or "amphitropical" distribution in the north and south temperate zones of the world (more about this later, p. 45), but even *Bembidion* diminishes in the extreme south, only 2 species reaching Puerto Williams (see Fig. 2).

Within the limits imposed by climate at nearly 55° S, Puerto Williams is a fine, diverse locality with much heavy forest, some open country, sphagnum bogs, small rivers, beaches along the Beagle Channel, and open subantarctic mountain tops. This whole locality has been well collected, but only 21 species of Carabidae have been found there in all habitats together. At least ten times as many species of Carabidae would be found in a similarly diverse area of the same size in the tropical part of South America. The

Puerto Williams carabid fauna is listed and analyzed in Table 3.

What I have said about Carabidae in southernmost South America is oversimplified but does illustrate how the dominant tropical tribes and genera of an important insect family decrease and dis-

TABLE 3. Carabid beetles at Puerto Williams, Navarino Island, Chile, *ca.* 55° S (from original data); see text for further details of distribution of Migadopini, Broscini, Trechini, and *Bembidion;* +, winged, −, wings atrophied, (±), wings dimorphic.

Species	Wings	Habitat	Taxonomic status	Geographic relationships
Ceroglossus suturalis F.	−	Forest floor	Genus endemic to S. S. Am.	Complex
Migadopini: 2 genera and species	− +	Forest floor	Genera endemic to S. S. Am.	Tribe also in S. Australia, N. Z.
Broscini: 3 species of *Cascellius*	− − −	Forest floor	Genus endemic to S. S. Am.	Complex (see text)
Trechini: 4 species of *Trechisibus*	− − − −	Open: on beach, and above timber line	Genus endemic to S. S. Am.	Subtribe also in S. Australia (not N. Z.)
Trechini: 1 species of *Homalodera*	+	Forest: subarboreal	Genus endemic to S. S. Am.	As above
Trechini: 1 species of *Kenodactylus*	−	Intertidal	Species endemic to S. S. Am.	Related sp. on islands S. of N. Z.
Bembidion: 2 species	+ (±)	1 on beach; 1 in sphagnum bogs	Species endemic to S. S. Am.	Genus mainly northern
Pterostichini: 1 species of *Argutoridius*	+	Open places	Genus endemic to S. S. Am.	Undetermined
Agonini: 1 species of *Agonum*	+	Sphagnum bogs	Species endemic to S. S. Am.	Genus mainly northern
Antarctiini: 4 species	+ + + +	1, forest floor; 3, open places	Tribe endemic to S. S. Am.	Undetermined
Plagiotelum irinum Sol.	+	Probably forest, subarboreal	Species endemic to S. S. Am.	Genus also in Tasmania (not N. Z.)

appear toward the south, most of them failing to reach extreme southern habitats, where their places are taken (in part) by a comparatively small number of species of other, special, southern cold-temperate tribes or genera. However, the details of distribution and relationships of the few Carabidae that do occur in extreme southern South America, for example at Puerto Williams (Table 3), are so diverse as to suggest a variety of different dispersal histories.

At the southern limits of their ranges some insects occur only in special habitats where they receive maximum warmth. For example, the one species of ant that I found at Puerto Williams, at the extreme southern limit of the known distribution of ants, was nesting on sphagnum bogs, actually on sphagnum humps well above water level. The bogs have no tree cover, and their surface therefore receives maximum insolation. The ant was nesting commonly among sphagnum strands near the surface of the humps, where the sphagnum was warm after even a few minutes of full sun. This ant is a rather small, undistinguished, Lasius-like species, *Lasiophanes picinus* (identified by E. O. Wilson). The two species of *Bembidion* at Puerto Williams (see third preceding paragraph, and Table 3) were in open, warmth-receiving habitats too, one on the sphagnum bogs and the other on the upper ocean beach above tide line. Both species had wider ecologic ranges near Punta Arenas.

The idea of climatic or ecologic barriers is a very old one, and "ecological zoogeographers" have been much concerned with the details of them recently, but I doubt if the profound effect of climate on animal distribution is often fully realized even now. Climate not only limits the distributions of many species and of many dominant higher taxa of animals, but by doing so makes protected places where other animals may survive. In the extreme south of South America, for example, in a small area including the southernmost forest and southernmost moorland, climate apparently bars out all frogs and lizards (and all other amphibians and reptiles), most or (in places) all ants, and most groups of predaceous carabid beetles that are dominant in the tropics. Frogs, lizards, ants, and Carabidae are probably four of the six most important groups of insect-eating animals in most parts of the world (the others being spiders and insectivorous birds), and their absence or fewness in the extreme southern corner of South America makes a place where predation and competition must be very much reduced. It is precisely here that special southern cold-temperate groups of insects are most concentrated, although some of them extend into less protected places too. Presumably they are forms that can tolerate the climate but are less tolerant of the predation or competition that they encounter elsewhere.

The climatic factor primarily responsible for the decrease in numbers of insects and other animals toward the southern tip of South America is apparently not killing cold but lack of warmth needed for activity and reproduction. This is consistent with the general fact that frogs, reptiles, and other warmth-demanding animals range farther south in the steppe zone than in the forest or moorland (see again Fig. 2). Extremes of cold as well as of heat may be greater on the comparatively dry, open steppe than in forest or on wet moorland at the same latitude, but the steppe receives much more total heat than the other zones, on which insolation is reduced by frequent cloud cover and rain. (See Landsberg *et al.* 1963, Map 3, for an oversimplified indication of the sunshine zones in southern South America.) The most heat-deficient habitats are therefore the wet forest and moorland, where the effect of latitude is increased by local climatic factors, and this is just where the plants and animals characteristic of the warmer parts of South America are fewest, and the special southern cold-temperate groups relatively most numerous. This generalization is supported by the fact, already noted, that some insects reach their extreme southern limits only in special, warmth-receiving habitats.

In southern Australia and Tasmania (Fig. 3) the diminution of the fauna is similar in general but much less extreme than in southern South America, for the land does not extend nearly so far south. The latitude of the southern tip of Tasmania is about $43\frac{1}{2}°$ S, and that of Cape Horn, 56° S. Nevertheless many groups of animals that are dominant and widely distributed in Australia are reduced or absent in Tasmania. For example, lizards are reduced in numbers, and the lizard families Varanidae and Gekkonidae, the latter a dominant insectivorous group in Australia, are absent in Tasmania (Lord and Scott 1924:110–111; one record of a *New Zealand* gekkonid in Tasmania is presumably an error). Turtles too are absent in Tasmania, as they are in southernmost South America. The ants of Tasmania are "a much-diminished extension" of the southeastern Australian ant fauna (W. L. Brown,

Jr., personal communication, 1963). The southeastern Australian state of Victoria has perhaps 300 species of ants; the whole of Tasmania, perhaps 150 (all belonging to Australian genera and most to Australian species); and ants are notably few in the cold, wet forests and moorlands of Tasmania, as they are in such habitats elsewhere in the world. Among Carabidae, the tiger beetles, which are widely distributed in Australia, are absent in Tasmania and not many species occur even in the southeastern corner of Australia (Sloane 1906:350). The large flightless Scaritini called "carenums," which are represented in Australia by about a dozen genera and hundreds of species, reach only northern and eastern Tasmania, and only three, nonendemic species occur even there (Sloane 1920). And *Tachys,* of which scores of diverse species occur in the warmer part of Australia, is very poorly represented in Tasmania. Several other dominant Australian groups especially of large carabids either do not reach Tasmania (in some cases not even the southern edge of Australia) or are poorly represented there, and a similar diminution of the fauna occurs in many other groups of insects. Lack or poverty of so many dominant Australian groups makes room in Tasmania for special southern cold-temperate groups of insects and other invertebrates.

In New Zealand, wide water barriers as well as climate limit the fauna and make place for special southern cold-temperate animals.

Further details are unnecessary. Enough has been said of existing situations, especially in southern South America, to show that, even without impassable physical barriers, climate profoundly limits and modifies floras and faunas in the far south.

Summary of effects of climate. The gross effect of climate on the distribution of plants and animals in the far south, notably in South America, can be summarized simply but superficially as a striking diminution of flora and fauna toward the south, with partial replacement of tropical groups by a comparatively small number of special southern cold-temperate ones, occurring mainly (but not exclusively) in south-temperate rain forest and on subantarctic moorland. But underneath this simple total effect of climate the details are diverse and complex. Climate affects different sorts of plants and animals differently. Striking replacements occur in some groups (fishes in fresh water, and some carabid beetles) but not in others (birds, mammals, and ants). Replacements often involve transitions rather than sudden changes. And the plants and animals concerned seem to have geographic relationships so diverse as to suggest that they have dispersed by different routes or at different times. Nevertheless, somehow, climate strictly limits the total size as well as the details of composition and distribution of the floras and faunas that contain all these diversities. The evolutionary implications of this effect of climate will be considered in Chapter 5.

The situation in South America shows that climate can make very great differences in biotas at opposite ends of one continuous piece of land. It follows that, whether now or in the past (in the fossil record), when plants or animals on different parts of what is now one continent are different, we do not necessarily have to divide the continent or imagine a physical barrier across it. Climatic differences may have produced the observed biotic differences.

The change of insects and other invertebrates as well as of plants from open steppe (in southern South America) or from open forest (in Tasmania) to closed rain forest is striking. The difference is so great as to suggest that the rain-forested areas (and moorlands) of southern South America and Tasmania, with New Zealand, might be recognized as a separate biogeographic region. However, there are three arguments against this. First, the areas in question are very limited, including (for example) only part of Tierra del Fuego and only part of Tasmania, and are distinguished ecologically rather than geographically. These areas are not really comparable, either singly or together, with other biogeographic regions. Second, terrestrial vertebrates, which are important animals biogeographically, do not conform to the pattern of the plants and invertebrates. And third, resemblances even of the invertebrate faunas of the wet forests of Chile and of Tasmania are really less than they seem at first. The Tasmanian rain forest is inhabited not only by many special southern cold-temperate insects with relatives in Chile but also by many other insects belonging to Australian groups that are absent in the South Chilean rain forest. Among Carabidae, for example, Pterostichini and Licinini are dominant in Tasmanian rain forest as they are everywhere in Australia (and Pterostichini are dominant over much of the world), but these tribes are absent in the southernmost rain forest in Chile. Under these circumstances the best course seems to be not to make a climatically or ecologically defined southern cold-temperate biogeographic "region" but just to stress the importance of climate and ecology in the far south and to try to find in what ways different climatic and ecologic factors affect the distributions of different southern plants and animals. (See, however, Kuschel (1964) for an interesting discussion of a tentative Austral Biogeographic Region, to be tested by biogeographers.)

3. Distribution around the
   southern cold-temperate zone

Southern beeches. The general pattern of distribution of many terrestrial plants and invertebrate animals in the colder part of the Southern Hemisphere is exemplified by the genus *Nothofagus,* the southern beeches. This is the only southern genus of the mainly northern family Fagaceae. Within the Fagaceae *Nothofagus* is considered most closely related to *Fagus,* the true beeches of the north, which occur in Europe, parts of eastern Asia with Japan and Formosa, and eastern North America with part of Mexico. Together, *Nothofagus* and *Fagus* form an amphitropical pattern of distribution, that is, a pattern of occurrence on both sides of the tropics (Fig. 5). The species of *Nothofagus* are evergreen or deciduous trees and shrubs. Different species of them form forests in all three principal land areas in the southern cold-temperate zone: southern

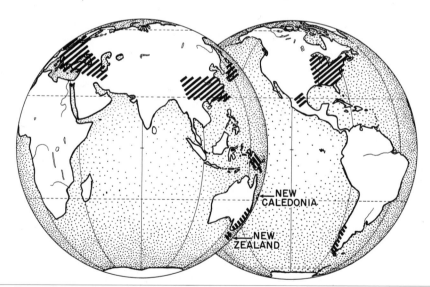

Fig. 5. Present distribution of southern beeches (*Nothofagus,* Southern Hemisphere including New Guinea) and true beeches (*Fagus,* Northern Hemisphere, details by courtesy of Dr. S. Y. Hu). Distributions are discontinuous in some of the areas indicated, especially on the mainland of Australia. See also Fig. 20.

South America, Tasmania with southeastern Australia, and New Zealand. This is a preliminary statement. It is true, but incomplete and oversimplified. It is, incidentally, all that Hooker and Darwin knew about the distribution of these plants.

Many other groups of plants and many of invertebrates have southern distributions that follow the *Nothofagus* pattern in a general way. However, details differ in almost every case, and the details are important and sometimes unexpected. In *Nothofagus* itself the details are as follows. In southern South America, trees of this genus form nearly continuous heavy forest on or near the western edge of land from central Chile (about 36° S) south to northern Navarino Island (about 55°, south of the Beagle Channel), and I saw patches of stunted evergreen *Nothofagus* in sheltered places still farther south, at Orange Bay (about 55½°), only about 50 km north of the latitude of Cape Horn. The northern end of the Chilean forest includes many additional genera of trees, but the southern end consists mainly of pure stands of single species of *Nothofagus*. In Tasmania, *Nothofagus* forms heavy forest at low altitudes in the southwest and forest patches on mountains elsewhere, and in southeastern Australia patches of it occur on isolated, very widely spaced plateaus and mountains from southwest of Melbourne (the Otway Ranges) east and north to the southern border of Queensland (the McPherson Range, at about 28° 20′ S and 1200 m altitude). In New Zealand, heavy *Nothofagus* forest occurs along much but not all of the western side of South Island, and tracts of *Nothofagus* are or were widely scattered elsewhere on both main islands of New Zealand, but the genus is absent on Stewart Island at the southern end of New Zealand. It is absent also on the Falkland Islands (which are treeless), east of the southern tip of South America, and it is in general absent on smaller and more remote islands.

It will be seen that, although *Nothofagus* is widely distributed in the south temperate zone, especially in the cooler areas, it is not ubiquitous there. Its occurrence elsewhere is relatively limited. However, what would otherwise be a consistent pattern of south-temperate zonation is complicated by an inconsistent detail, by occurrence of the genus also on New Caledonia and New Guinea. The extent of occurrence of *Nothofagus* on New Guinea, almost on the equator, was not realized until about ten years ago (van Steenis 1953), although the genus was first discovered on the island in

1913. It forms extensive but patchy stands on mountains there, especially in the lower montane rain forest between about 2300 and 2800 m, with extreme altitudinal limits of about 1000 and 3000 m (van Steenis 1953; Robbins 1961). That *Nothofagus* can exist and reproduce at 1000 m in tropical New Guinea suggests that, in spite of its occurrence mainly in the southern cold-temperate zone, the genus is cold-tolerant rather than cold-adapted and that it might disperse through tropical or at least subtropical areas.

The fact that the south-temperate zonation of *Nothofagus* is complicated by an inconsistent detail is important. The distributions of many plants and animals that are characteristic chiefly of southern cold-temperate areas are complicated by unexpected details of one sort or another. These inconsistencies should always be looked for and emphasized. They are likely to be significant.

All this about *Nothofagus* is well known, but I can add one point of interest. I have collected at the *Nothofagus* level on the Bismarck Range in Northeast New Guinea, and found no antarctic or southern cold-temperate Carabidae there, and none was found at similar levels on the Snow Mountains of West New Guinea by Toxopeus, whose collections I have seen, and who made a fine collection of high-altitude Carabidae during the Third Archbold Expedition. The montane forest of New Guinea does have a rich fauna of Carabidae, but all or almost all of them are derived from the surrounding tropical lowlands, not from distant temperate areas. Some other plants besides *Nothofagus* on the mountains of New Guinea do belong to south-temperate groups (van Steenis 1953), but they seem to be more or less independent elements, not members of a south-temperate biota isolated as a whole on New Guinea.

The fossil record and geographic history of *Nothofagus,* with some additional details of the present distribution of groups, will be considered in Chapter 16.

Peloridiid bugs. The family Peloridiidae is a small family of primitive Hemiptera-like Homoptera. The adult insects (Fig. 6) are small, usually between 2.5 and 5 mm long, flattened, with the sides of the body expanded and with conspicuous reticulate venation. Ten genera and 19 species of the family are known. They are all short-winged and therefore flightless, with one exception. The exception, *Peloridium hammoniorum* of southern South America, is now known to be dimorphic. Both long-winged and short-winged in-

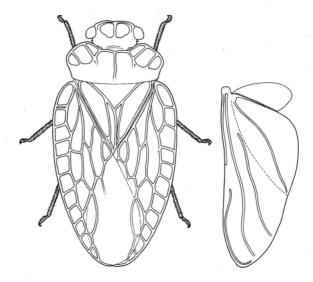

Fig. 6. A peloridiid bug, *Peloridium hammoniorum* Breddin, of southernmost South America (after China 1962). The insect is 4.7 mm long and its inner wing, shown detached, is drawn to the same scale as the whole insect.

dividuals occur in both sexes, and long-winged ones fly. On Navarino Island they have been found on snow on mountains, where they had evidently fallen after flight. All species of the family, so far as is known, spend the active part of their life in wet moss in humid forest, but they apparently shelter in leaf mold on the forest floor during dry periods and sometimes persist for a long time after forests have been felled, breeding in damp moss which grows on the ground in the wet season. They apparently feed on the moss itself. This brief summary of what is known of the biology of peloridiids is chiefly from a recent paper by China (1962:136, 142, 143).

The geographic distribution of the family Peloridiidae (Fig. 7) is like that of *Nothofagus* in general, but not in all details. The peloridiids, like *Nothofagus*, occur principally in southern South America, Tasmania and southeastern Australia, and New Zealand. Details (chiefly from China's paper) are as follows. In South America, 4 genera, 6 species are now known in southern Chile and adjacent parts of Argentina. Extreme northern and southern localities apparently are in the vicinity of Valdivia (about 40° S) and on Navarino Island (about 55° S). In Tasmania and at widely spaced localities on the mainland of southeastern Australia there are 3 genera, 5 species. Two of the genera, with 4 species, are confined to Tasmania and southern Victoria. The third, monotypic genus occurs on the McPherson Range on the New South Wales–Queensland border (at about 28° 20′ S and 1200 m altitude) and elsewhere at widely scattered points at comparable altitudes in northern New South Wales (T. E. Woodward, personal communication, 1963). In New Zealand there are 2 genera, 6 species, scattered over both main islands and Stewart Island.

The distribution of peloridiids as described thus far is a consistent one of occurrence on the three principal pieces of land in the southern cold-temperate zone, with slight occurrences northward in western South America and eastern Australia. However, as in so many other groups of far-southern organisms, the distribution

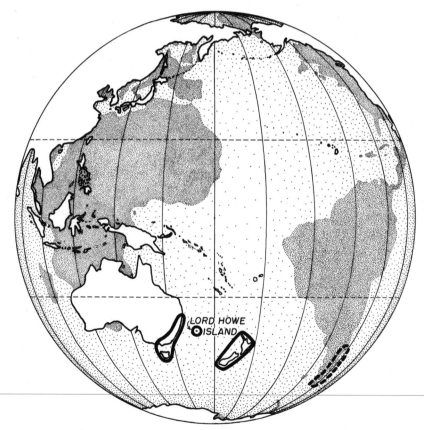

Fig. 7. Distribution of bugs of the family Peloridiidae. The distribution on the mainland of Australia is actually very discontinuous within the area indicated.

includes an additional, inconsistent detail. In this case the inconsistent detail is the occurrence of an endemic genus with 2 species on Lord Howe Island, an apparently strictly oceanic (volcanic) island about 500 km (300 mi) east of Australia at 31° 30′ S. The peloridiids on Lord Howe Island apparently live only in permanently wet moss in a few acres of forest on the summit plateau of Mount Gower, at slightly less than 1000 m altitude (Evans 1959).

Peloridiids often occur in *Nothofagus* forest, and in Australia these insects and *Nothofagus* reach the same northern limit, on the McPherson Range. However, the peloridiids do not depend on *Nothofagus,* for the latter is absent on Stewart Island and Lord Howe Island, where peloridiids occur.

It should be added that the distribution of peloridiids is still probably not completely known. They are obscure insects, usually found only by special collecting methods. They would probably not be known on Lord Howe Island except for a virtual accident, the accident being that Lea happened to get one nymph there in general collecting about half a century ago. It was the existence of this single old specimen that led John Evans to make his successful search of the island for peloridiids.

The geographic history (history of dispersal) of peloridiids is not known, but it seems safe to say that they are an ancient group and that they probably have been in the far south for a long time. They may be derived from, or related to, the fossil family Ipsviciidae of the Upper Triassic of northeastern Australia (see again China 1962:159). Also, their place in phylogeny, as very primitive Hemiptera-Homoptera, suggests that they are an old group. And also, the fact that all the genera of southern South America, Australia-Tasmania, and New Zealand are different suggests an old dispersal. The means of dispersal of the insects are more doubtful. It used to be thought that they could not fly and could disperse only through continuous wet forest over continuous land, and that their presence in widely separated places in the southern cold-temperate zone required direct land connections. However, the fact that peloridiids sometimes persist for considerable times after forest is cut suggests that their requirements are not as strict as was formerly supposed. And the fact that long-winged individuals occur in at least one species, and do fly, shows that peloridiids are derived from winged ancestors and suggests that the ancestral forms may have dispersed by flight. They seem to be just the sort of insects most likely to be carried long distances by wind. Their small size favors wind dispersal. They evidently fly well enough to be exposed to wind, but not strongly enough to control their movements afterward. The presence of peloridiids on oceanic Lord Howe Island suggests—I think almost proves—that the insects have sometimes crossed considerable water gaps.

Attempts have been made to deduce the place of origin and directions of dispersal of peloridiids from their present distribution and from the nature (relative primitiveness and the like) of different genera in different places. However, I think the deductions are based on doubtful premises. I think all that can safely be said of the geographic history of these insects is that they probably dispersed long ago, perhaps in the Mesozoic; that they have probably lived in the cooler part of the Southern Hemisphere for a long time; and that they have probably evolved and dispersed complexly in the far south, although the details of their dispersal are undecipherable at present. Or perhaps it would be better to say that the history of peloridiids cannot now be deciphered from their own evidence, but that they will probably fit into the general geographic history of life in the far south when the history is known from other evidence.

Migadopine carabid beetles. The tribe Migadopini is a well-defined tribe of medium-sized Carabidae (Fig. 8), characterized by extension of the scutellar striae nearly to the apex of the elytra and by other technical characters. These beetles live on the ground (except that the Australian *Decogmus* may occur on tree trunks) mostly in south-temperate rain forest or on subantarctic moorland.

Migadopini occur in three general areas (Fig. 9): southern South America and associated islands including the Falkland Islands (7 genera, some localized in small parts of this general area), southern and western Tasmania with southeastern Australia (4 genera, 2 of them localized at different places on the Australian mainland), and New Zealand with the Auckland Islands (4 genera). All the genera have restricted ranges: none occurs in more than one of the three general areas indicated. Migadopines do not occur anywhere else in the world, so far as is known. The tribe as a whole therefore has a strictly south-temperate zonal distribution, and most of the genera occur only in cool or cold places in the south (but see below).

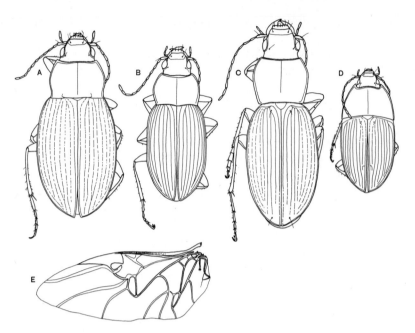

FIG. 8. Carabid beetles of the tribe Migadopini: *A, Migadops latus* Guér. of southern South America; *B, Antarctonomus complanatus* Blanch. of southern South America; *C, Calyptogonia atra* Sl. of Tasmania; *D, Stichonotus piceus* Sl. of Tasmania; *E,* wing of *Antarctonomus complanatus.* All ca. $4\frac{1}{2}\times$.

Jeannel, in his useful revision of the Migadopini (1938), considered the tribe wholly flightless (apterous) and thought it had originated on an ancient antarctic continent and spread from there. However, as usual in "antarctic" distributions, there are inconsistent details. First, the tribe is not wholly flightless. Both *Antarctonomus* in Chile and *Decogmus* in eastern Australia are strongly winged. Second, the tribe is not quite confined to *cold* areas in the south. The northernmost genera in both South America and Australia, although technically within the south temperate zone, occur at warm-temperate or subtropical localities: *Rhytidognathus* is known only from Uruguay, about 35° S; *Decogmus,* from Comboyne in northern New South Wales, at 31° 36′ S and less than 1000 m altitude. Third, existing genera of the tribe are extraordinarily diverse in form and characters, as if they are products of a complex ecologic as well as geographic radiation rather than of simple spread from an antarctic center. And fourth, the closest relatives of the tribe are probably the Elaphrini of the *north* temperate zone, although the relationship may not be very close (Jeannel 1938; Lindroth, personal communication, 1963). All this suggests that the ancestor of the Migadopini was winged, that it may have lived in or dispersed through relatively warm climates, that the history of the tribe has been complex, and that a common ancestor of this tribe and the Elaphrini crossed the tropics a long time ago. These details do not disprove an antarctic origin of the Migadopini but do suggest other possibilities.

Other plants and animals. Many other plants and many invertebrate animals have southern distributions that in general (but usually not in all details) resemble the distributions of *Nothofagus*, Peloridiidae, and Migadopini. The principal examples among plants are listed by van Steenis (1962, especially pp. 256–265, maps 10–15). Other examples will be found in papers cited in the Bibliography,

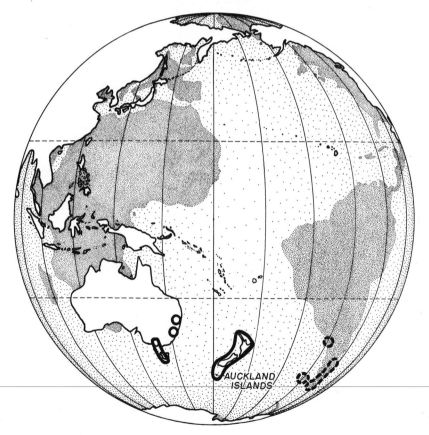

FIG. 9. Distribution of carabid beetles of the tribe Migadopini.

and additional examples from carabid beetles are given in the following pages. But this pattern of distribution in the far south does not occur among terrestrial or strictly fresh-water vertebrates. South America and Australia do share some notable groups of vertebrates, including marsupials, leptodactylid and hylid frogs, and chelyid turtles, but these are all mainly tropical or warm-temperate animals. Most of them do not reach the southern cold-temperate zone at all, and those that do reach it do not show special relationships between the southern cold-temperate forms on different continents. This fact is less well known but very significant. I shall come back to it in Chapter 6.

Only certain small fishes, especially the family Galaxiidae, come close to being an exception to this fact. Galaxiids do occur chiefly in fresh water in the southern cold-temperate zone: in southern South America, southern Australia and Tasmania, and New Zealand. (They occur also on the southern tip of Africa and on New Caledonia.) However, some of these fishes enter or breed in the sea and may have dispersed through it, or their ancestors may have done so. This possibility has been known and argued about for more than half a century and need not be re-argued here (see references in Darlington 1957:71, 72, 107; and see also McDowall 1964).

4. Southern distributions in relation to the world: four groups of carabid beetles

To illustrate the fourth and fifth principles listed at the beginning of Chapter 1, that situations in the southern cold-temperate zone should be considered in relation to the whole world and that it is important not only to investigate single cases in detail but also to compare cases, I shall take a set of four groups of Carabidae.

Migadopini. The distribution of this tribe is described in sufficient detail in the preceding chapter. The tribe itself is confined to the south temperate zone and most of the species are southern *cold-*temperate or subantarctic in distribution (Fig. 9). However, the closest relatives of the Migadopini are apparently the Elaphrini, and the latter are confined to the cooler parts of the north temperate zone, in temperate Eurasia and North America. This suggests (as one possibility but not the only one) the existence a long time ago, perhaps in the Mesozoic, of a common ancestral stock that was adapted primarily to temperate climates but that somehow crossed the tropics, in one direction or the other, and then evolved separate tribes in northern and southern temperate areas.

Broscini. Occurring with Migadopini in the wet forests of south-western South America and Tasmania are certain members of the tribe Broscini, a tribe of medium-sized or large, ground-living Carabidae (Fig. 10*A*). Some of the South Chilean and Tasmanian forest-living forms are very much alike, although they belong to different genera, *Cascellius* in Chile and *Promecoderus* in Tasmania. These two genera may actually be related, but probably less closely than their external similarity suggests. Certain particular species of these two genera resemble each other remarkably, perhaps as a result of convergence in similar niches in the South Chilean and Tasmanian forests. For example, *Cascellius* in Chile and *Promecoderus* in Tasmania each include, living in wet *Nothofagus* forest, a medium-sized bright-green species and a smaller brown one. An observer who knew only the Chilean and Tasmanian faunas might be overimpressed by these similarities.

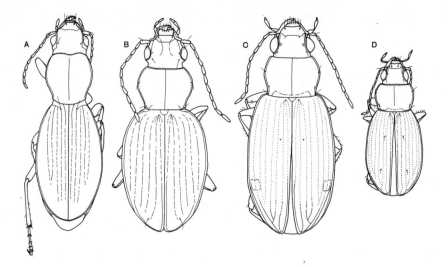

Fig. 10. Subantarctic carabid beetles from Puerto Williams, South Chile: *A, Cascellius gravesi* Curt. (Broscini); *B, Trechisibus antarcticus* Dej. (Trechini); *C, Bembidion kuscheli* Jean.; *D, Bembidion nitidum* Jean.; *A,* ca. 4.2×; others, ca. 10×.

However, other broscines exist, and their distribution as a whole is amphitropical (Fig. 11). They occur around the world both north and south of the tropics but are absent in the tropics themselves, except that primarily temperate genera apparently enter the edges of the tropics in Asia and Australia. The geographic relationships of broscine genera have been discussed several times recently (Jeannel 1941; Britton 1949; Ball 1956). I shall be concerned here mainly with the southern forms. Of the northern ones I need say only that they do exist and that some of them are winged, as the ancestor of the tribe must have been. The wings have atrophied in all the southern forms, so far as I can determine.

In the Southern Hemisphere, Broscini occur in the same three general areas as Migadopini: southern South America, southern Australia with Tasmania, and New Zealand. However, details are different. Broscines are more widely distributed than migadopines in open steppe as well as forest in southern and western South America and in open as well as forested parts of southern Australia, while migadopines are more widely distributed on the southernmost moorland. For example, migadopines are represented by endemic genera on subantarctic moorland in extreme southern South America and the Falkland Islands and also on the Auckland Islands south of New Zealand, while broscines are undifferentiated on the southernmost moorland in South America and absent on the islands in question. In South Chile several species of broscines of the genus *Cascellius* do extend onto subantarctic moorland, but they are all primarily forest-living species and the genus as a whole is primarily forest-living.

On the whole, although Migadopini and Broscini occur together in many places, the distribution of the Migadopini within the south-temperate zone is more southern than that of the Broscini, and the Migadopini are more differentiated in the southernmost habitats. This difference suggests that migadopines have been in the southern cold-temperate zone longer than broscines,

FIG. 11. Distribution of carabid beetles of the tribe Broscini. Areas of occurrence in the Southern Hemisphere are bounded (except that northern limits in Australia and South America are not determined), and southern limits in the Northern Hemisphere are shown (approximately) by heavy lines. Double-ended arrows indicate the geographic relationships of the broscids of different areas.

long enough for generic differentiation on the moorland, while
the broscines have moved south more recently and are just beginning to invade the southernmost moorland. However, broscines
have evidently been in the Southern Hemisphere for a considerable time. All the southern genera are different from northern
ones, and no genus occurs in more than one of the three main
areas of distribution in the south.

Several genera of Broscini occur on New Zealand. They are
all endemic and all may be derived from a single ancestor (Ball
1956:44, 47). This hypothetical ancestor of all New Zealand Broscini presumably reached New Zealand from Australia. Related
forms still occur in Australia, and also in the Northern Hemisphere. However, the New Zealand broscines are *not* directly related to the small, forest-living forms (*Cascellius* and *Promecoderus*)
that are so similar in Tasmania and southern Chile. The result
is that, while there are or may be direct relationships between
some Australian and South American broscine genera, and between other Australian and New Zealand genera, there are no
direct relationships between the broscines of South America and
New Zealand (see arrows on Fig. 11). If the existing Australian
relatives of the New Zealand broscines should become extinct,
some surviving Australian and South American genera would be
(or seem to be) related, while the New Zealand forms would
seem to be derived from a separate, northern stock. This pattern
of relationship does occur in some other groups of Carabidae,
including the Trechini (below; and see p. 64).

Trechini. A third tribe of Carabidae important in the southern
hemisphere is the Trechini. It is a large tribe of small carabids
(Fig. 10*B*). Some are winged; others are flightless, with atrophied
wings. Some of the winged forms live beside running water. Most
of the flightless ones live on the ground in forest, on moorland,
on mountains, or in caves. A few trechines live on the seashore
between tide lines and may disperse on ocean drift, but they
form a separate taxonomic group and are not concerned in the
history of the terrestrial members of the tribe.

Jeannel (1926–1928) has published a classic monograph of the
tribe Trechini (which he considered a subfamily). The tribe as
a whole is almost amphitropical in distribution (Fig. 12), best
represented in the north temperate and south temperate zones,

Fig. 12. Distribution of carabid beetles of the two principal terrestrial subtribes of the tribe Trechini. Areas of occurrence in the Southern Hemisphere are bounded (the occurrence in southwestern Australia is based on the discovery of an undescribed species there), and southern limits in the Northern Hemisphere are shown (approximately) by heavy lines. Triangles indicate geographic relationships: Homaloderina occur in Australia (but not New Zealand), South America, and Spain; Trechina, in the Northern Hemisphere and New Zealand.

although a few water-loving genera occur in or across the tropics. However, the tropical forms are mostly winged. Flightless trechines are almost all either northern or southern, not tropical. Most of the flightless species belong to two principal subtribes (called tribes by Jeannel), one mainly northern, the other mainly southern. Although most species of both subtribes are now flightless, both include some winged species and both are evidently derived from winged ancestors presumably capable of dispersal by flight. The northern subtribe (Trechina) is dominant across temperate

Eurasia and North America. This subtribe is absent in South America and in Australia and Tasmania but is represented on New Zealand by a group of 3 interrelated endemic genera (all, I should think, derived from one ancestor) and at least 9 species, most of them described by Britton (1962 and other papers). The other principal subtribe (Homaloderina) includes most of the terrestrial trechines of southern South America and southern Australia and Tasmania, and they occur in the southernmost wet forest and on moorland in both places. However, the South American and Australian-Tasmanian genera are all different. And this subtribe is not represented on New Zealand but is represented in the Northern Hemisphere by one blind, presumably relict, monotypic genus in a cave in Spain! This pattern of distribution is indicated by triangles in Fig. 12. The geographic pattern is so surprising that I doubted the reality of it until I had personally confirmed the characters on which the classification is based.

The classification of subtribes of Trechini is based on structural characters including mandibular teeth, and the characters do seem to be valid. The structural classification is partly but not wholly supported by an apparent difference in adaptive potential of the subtribes. Members of the northern subtribe commonly enter caves and become blind and otherwise highly modified there (in Europe, Japan, eastern North America, and elsewhere), and some of the supposedly related New Zealand forms are cave dwelling too (Britton 1962). On the other hand, the southern subtribe has apparently produced no true cave dwellers in South America or Australia-Tasmania, while the supposedly related relict in Spain is a blind cave dweller, and this fact may throw some doubt on its supposed relationship.

However, regardless of details, the tribe Trechini (Jeannel's subfamily Trechinae) surely has two principal zones of dominance, north and south of the tropics, and several different members of the tribe surely have crossed or are crossing the tropics, in one direction or the other. Each of the two principal subtribes discussed above has crossed at least once, and three additional, primarily winged groups extend part or all of the way across the tropics now. Those that do so in the Old World are *Perileptus,* which ranges almost continuously from Japan and southeastern Asia to southeastern Australia (and from Europe to South Africa), and *Trechodes,* which occurs very discontinuously in southeastern

Asia, the Philippines, and Australia (and in Africa and Madagascar), with a related genus, *Cyphotrechodes,* on mountain moorland in Tasmania. And, in the New World, trechines of the genus *Cnides* range from part of Central America to part of Chile.

Bembidion. The fourth and last group of Carabidae to be discussed now is the genus *Bembidion.* This is a very large genus of small carabids (Fig. 10*C, D*) which live on the ground and are comparable to trechines in size and habits, except that *Bembidion* rarely enters caves. Some *Bembidion* are winged and some have atrophied wings, and some live beside water and others away from it. A few live on the seashore or between tide lines, but these are not concerned in the distribution of the dominant terrestrial forms. The genus *Bembidion* as a whole has not been revised, but fortunately Jeannel (1962) has recently reviewed the known South American species, and I (1962), the Australian ones. The Eurasian ones have been put in basic order by Netolitzky (1942–1943) and most of the North American ones by Lindroth (1963a). Netolitzky, Lindroth, and I have treated *Bembidion* as one large genus with many subgenera, while Jeannel has split it into many separate genera, but this difference in usage does not affect the distribution of the group and need not concern present readers. In any case, *Bembidion* in the broad sense is a much more compact group than the tribe Trechini and is therefore probably more recent in evolution and dispersal.

The distribution of this single genus (Fig. 13) roughly parallels that of the whole tribe Trechini. *Bembidion,* like the Trechini, is bizonally dominant, with hundreds of species around the world north of the tropics in Eurasia and North America, and also fair numbers in southern South America, a few in southern Australia and Tasmania, and some on New Zealand. The genus is very poorly represented in the tropics, although a few species do occur there in both Old and New Worlds. Most of the few tropical species live at high altitudes on mountains or, if at low altitudes, beside fresh or salt water, the presence of which seems to give some protection against tropical climate. Most of the few lowland tropical species are winged as well as water-loving. For some further discussion of the general distribution of *Bembidion* see Darlington 1953 and 1959c.

The distribution of *Bembidion* in southern South America has

Fɪɢ. 13. Distribution of carabid beetles of the genus *Bembidion*. Terrestrial forms occur on all habitable continents and principal islands of the world, except New Guinea, but are concentrated principally north and south of the tropics, in areas indicated by heavier hatching. Arrows show apparent direction of dispersal of 7 stocks of the genus into the Southern Hemisphere.

been noted in the discussion of southward diminution of the insect fauna (p. 22). About 70 species of the genus are known in South America, most of them along the Andes or at low altitudes in the warmer part of the south temperate zone. Some (all?) of them are related to and apparently derived from North American forms. In fact, the South American species of *Bembidion* include at least 3 northern subgenera (called genera by Jeannel), indicating at least 3 crossings of the American tropics by different members of the genus. At least 8 species range south to the Straits of Magellan on the opener, warmer, eastern side of South America, but only 2 reach the south side of the Beagle Channel, and they occur there only in special habitats that receive full sun (see p. 24). No species of *Bembidion* occurs in the southernmost forest or moorland in South America, although Trechini do so. However, one of the two southernmost South American species of *Bembidion* has entered another habitat characteristic of the far south, sphagnum bogs, and has lost or is losing its wings there. I have short-winged specimens of it from Puerto Williams.

Five native species of *Bembidion* occur in Australia, chiefly in the southern part of the continent (Darlington 1962). They are not directly related to South American forms but seem to have been derived independently from the north, from Eurasia or the Old World tropics, by three successive invasions of Australia. Four of the five species extend to Tasmania and, although three of them occur at only low altitudes there, the fourth reaches mountain moorland. Also, one of the species (not the one that reaches the mountain moors) is undergoing wing atrophy: most Tasmanian specimens of it are short-winged. There is thus a remarkable convergence between unrelated stocks of *Bembidion* in southernmost South America and Tasmania. In each place members of the genus, independently derived from the north, are now beginning to invade far-southern habitats and to evolve flightless forms.

The *Bembidion* of New Zealand have not been critically studied, but superficial examination suggests that they are derived not from any existing Australian or South American stock but from a separate ancestor which, if it came by way of Australia, has disappeared there. New Zealand possesses more, and more diverse, species of *Bembidion* than Australia does. This suggests a longer period of evolution in New Zealand, which is consistent with derivation of the New Zealand forms from a separate, older ancestor.

Apparent dispersal cycle of Carabidae. The geographic history of southern Carabidae must be deduced from their present distributions and from various indirect clues, for pertinent fossils are lacking.

When the four groups of Carabidae discussed above are compared in reverse order—*Bembidion* first and Migadopini last—it is seen that they may represent successive stages in a common, world-wide cycle of evolution and dispersal. The apparent cycle is: rise on the large land masses in the Northern Hemisphere, or possibly in the tropics; dispersal southward into southern South America and southern Australia by separate routes, and to New Zealand probably from Australia; disappearance of the tropical or tropics-crossing forms, leaving an amphitropical pattern; and finally disappearance from the Northern Hemisphere, leaving survivors on the three main pieces of land in the southern cold-temperate zone. (Other groups might disappear in the Southern Hemisphere and survive only in the Northern.)

All four groups of Carabidae discussed above still include winged forms. All the groups are therefore derived from winged ancestors probably capable of dispersing rapidly, for long distances, and across barriers. All four groups can tolerate cold climates (some Carabidae apparently cannot), and all now occur principally in cool- or cold-temperate areas. But all four groups have also a few representatives either actually in the tropics (*Bembidion* and Trechini) or in warm-temperate parts of southern South America and Australia (Migadopini and Broscini). All the groups are therefore apparently cold-tolerant rather than strictly cold-adapted. Moreover, the tropical forms, when they exist, are usually winged and presumably able to make the long, multiple dispersals required by the hypothetical cycle. In most cases the existing tropical forms cannot be the direct ancestors of existing south-temperate groups, but they show how the latters' ancestors may have crossed the tropics.

If these four groups of Carabidae do represent stages in a common cycle of dispersal, *Bembidion* is just beginning it, or is nearest the beginning. This is suggested by the relatively slight differentiation of the genus in different parts of the world. It is neither well represented nor much differentiated in the *colder* parts of the south temperate zone. It is apparently just beginning to invade and become adapted to far-southern habitats, and is doing so, independently, in both southern South America and Tasmania.

The Trechini seem to have moved somewhat further than *Bembidion* through the hypothetical cycle. The dominant tribes have formed an essentially bizonal pattern of distribution in the temperate zones, although a few winged forms still extend across the tropics. Also, the trechines as a group have diversified much more than *Bembidion*, and have invaded, differentiated in, and radiated in far-southern habitats much more than *Bembidion* has done.

The Broscini now have a strictly bizonal distribution. They

must have crossed the tropics, but the tropics crossers have disappeared. To this extent at least Broscini have gone further than
Trechini through the hypothetical cycle.

Finally the Migadopini are nearest the end of the cycle. If
they ever existed in the tropics and the Northern Hemisphere,
they have disappeared there and survive as relics only in the

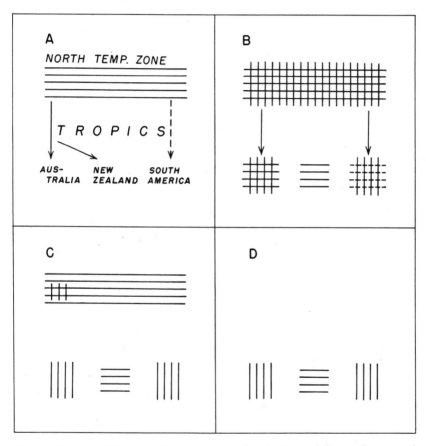

FIG. 14. Diagram of the hypothetical pattern of evolution and dispersal suggested
by carabid beetles. *A,* an initial stock evolves in the north and disperses southward to Australia, New Zealand, and perhaps South America. *B,* a second stock
evolves in the north and disperses southward to Australia and South America,
but not New Zealand. *C,* the *first* stock dies out in Australia and (if it existed
there) in South America, and the *second* stock dies out in the north except for
local relics. *D,* both stocks disappear completely in the north, leaving relics in
the three principal areas in the southern cold-temperate zone. Note that *C* is the
actual pattern of distribution of the principal subtribes of Trechini now (Fig. 12).
Many variations of this general pattern are possible.

south temperate zone. The differentiation, localization, and adaptation of different genera to different far-southern habitats (indicating long residence in the south) and the fact that their relatives, so far as they have relatives, are northern and tribally distinct (suggesting a crossing of the tropics a long time ago) are at least consistent with the Migadopini having passed through the suggested dispersal cycle long ago. But even if they have done so, they may have undergone further dispersal, later, in the far south.

The hypothetical dispersal pattern suggested by Carabidae is diagrammed in Fig. 14, and the complexity of it is further discussed, in connection with New Zealand, in Chapter 6.

PART II   The geographic patterns in
    relation to evolution and dispersal

5. Area, climate, number of species, evolution, and dispersal

The world-wide pattern of dispersal suggested by carabid beetles (Chapter 4) may be a product of the relation of evolution to area and climate. This hypothesis is so important to understanding the distribution of plants and animals in the far south that it must be reviewed here.

Area and number of species. Part of the factual basis of this hypothesis is the observed relation of number of species of plants and animals to area and to climate. A relation between area and number of species is suggested by comparison of almost any large continental island, say Great Britain or New Guinea or Tasmania, with a climatically similar but very small island, an acre or two in extent. If both islands have had the same history—if both have been separated from the same continent recently—the large island will have many more species of plants and animals. For example, the large island will probably have many species of mammals including large herbivores and carnivores, while the small island may have no terrestrial mammals or only one or two species of mice. This difference is so obvious and commonplace that most persons do not think about its evolutionary significance.

The limiting effect of area is not imposed abruptly. This fact is obvious too. There is no critical size above which islands support whole continental faunas and below which the number of species is suddenly drastically reduced. The number of species decreases in some sort of relation to decrease of area. For example, the number of species of amphibians and reptiles on different islands in the West Indies decreases with area (Table 4), and so does the number of species of ants on different islands in the western Pacific (Table 5). Table 4 suggests that, in the West Indies, division of area by 10 divides the number of species of amphibians and reptiles by 2, approximately. However, this ratio

TABLE 4. Relation of area to number
of species of amphibians and reptiles
on some islands in the West Indies
(from Darlington 1957).

Approximate area (sq. mi.)	Number of Species	
	Round	Actual
40,000	80	76–84
4,000	40	39–40
(400)	(20)	—
40	10	9
4	5	5

does not hold in all cases. The apparent relation of area to number of species has been formulated mathematically by Wilson (1961) and by other persons (for example, Hamilton, Barth, and Rubinoff 1964), but mathematical treatment is unnecessary here. The general relation between area and number of species is observable without mathematics and is fundamental. Area does somehow determine or limit number of species. And this is true on all land masses, from the largest continents to the smallest islands, although the effect of area is often complicated by other factors.

Climate and number of species. A relation between climate and number of species is obvious and fundamental too. Climatically favorable areas support many species; unfavorable ones, relatively few. Because climate is not a single factor but a combination of

TABLE 5. Relation of area to number of species of ponerine ants on some islands in the western Pacific (from original data from E. O. Wilson).

Islands	Approximate area (sq. mi.)	Number of species
New Guinea	307,000	112
Solomons	16,000	37
Fiji	7,000	17
New Hebrides	5,700	11
Samoa	1,200	6
Tonga	250	3
Rotuma	14	1

factors that are measured in different ways (by degrees of temperature, inches of rainfall, and so forth), the relation of climate to number of species is difficult to express in figures, but it can be illustrated. It is illustrated in the Northern Hemisphere, for example, by reduction in the number of species of trees toward the northern limit of forest, reduction of the amphibian fauna to one species of frog in arctic North America, and reduction of the mammal fauna to 8 species at the northern limit of land in both Eurasia and North America. In the Southern Hemisphere the effect of climate on number of species is illustrated by the diminution of flora and fauna from the tropical part of South America southward to the antarctic. The South American tropics support thousands of species of plants of diverse groups and many diverse animals too, including thousands of species of insects of many orders. For example, the Amazonian rain forest probably includes at least 2500 species of trees (Richards 1952:229), but only 6 species occur on Tierra del Fuego (p. 19), and none on Antarctica. The latter supports only 3 species even of smaller flowering plants, and they are confined to the Antarctic Peninsula. The only plants that now tolerate the climate of the main part of the Antarctic Continent are mosses, lichens, and algae. Terrestrial vertebrates diminish in numbers southward in South America (Chapter 2), and no strictly terrestrial vertebrate now occurs on Antarctica. Of insects (other than ectoparasites of birds and marine mammals), only two species of flies reach Antarctica, and they too are confined to the Peninsula or islands off the Peninsula, although Collembola and free-living mites reach parts of the Antarctic Continent proper (see Chapter 12). The decrease of number of species of land arthropods from the Auckland Islands to Campbell Island, Macquarie Island, and Antarctica is graphed by Gressitt (1964c:564, Fig. 10).

There is no one boundary between the tropics and the antarctic where tropical numbers and diversity suddenly stop and antarctic poverty suddenly begins. Plants and animals decrease in numbers southward in an irregular gradient. Some details of the decrease of flora and fauna southward in South America are given in Chapter 2. Tables 1 and 2, showing decrease southward of snakes and ants respectively, illustrate the kind of gradient that evidently exists in many other groups. It is legitimate to think of this gradient as extending to Antarctica, for antarctic

life is evidently limited primarily by climate rather than by the
water barrier: the fossil record of plants shows that the antarctic
flora has in fact been *reduced* to its present level, presumably by
Pleistocene cold and glaciation (Chapter 12).

All this about climate is necessarily oversimplified. Some special
groups of plants and animals evidently find temperate climates more
favorable than the tropics, and these groups may form gradients in
which numbers of species are greatest in temperate areas and decrease toward the tropics as well as toward the arctic or the antarctic, although the mass and diversity of terrestrial life as a whole
is surely greatest in the tropics. I cannot take space for further discussion of the almost endless complexities of this subject but can
only stress that an obvious and fundamental relation does exist between favorableness of climate and number of species in spite of
very great irregularities and complexities.

*Competition and selection in limited areas and adverse climates, and else-
where.* If a team of ecologists were to study the situation in southern South America in detail, they might be able to find the
special factors that now determine the southern limit of each
separate species of plant and animal, and this would be well
worth doing. But such a study would leave unknown two general
factors or processes that affect the situation. One is the effect of
evolution in the course of time, in changing the tolerances and
geographic limits of single species. And the other is the effect of
limited area and unfavorable climate on far-southern floras and
faunas as wholes—on total number as well as nature of species.

The hypothetical team of ecologists in southern South America
might find that the species now living in any given area and climate are the only ones within reach that are fitted to live there.
But it can hardly be supposed that these species just happen to be
preadapted to the situation and that the number of suitable preadapted species is, by chance, proportional to area and climate in
each case. Adequate forces or processes presumably limit the number of species that enter or evolve in a given area and climate. The
factors concerned may be hard to detect and may operate only occasionally. They are comparable to density-dependent factors in
populations, which act only when individuals become too crowded,
but which most biologists think must act at least occasionally to
keep populations of single species within limits. The density-depend-

ent factors that limit number of species apparently must involve competition and selection among species. No other biological mechanism is adequate to do it. Ecologists and mathematicians are beginning to investigate the factors that do determine number and diversity of species in different places, but their (preliminary) ideas and findings cannot be discussed in detail here. Future contributions to the subject are worth watching for, in the pages of *Evolution* and elsewhere.

The observed relation of number of species to area and climate thus apparently implies and requires an all-pervading, effective process of competition and selection *among species*. This process can, of course, be deduced from other evidence, for example from the general level and balance of all principal past and present faunas, each made up of large and small forms and herbivores and carnivores in reasonable proportions (Darlington 1957:552–553), and from the successions of groups of plants and animals that appear in the fossil record. The process of competition and selection among species must go on everywhere. The whole world is in fact a limited area, and the number of species that can occur on it has some limit even though the number is huge, and the limiting process must involve competition and selection. Darwin understood this as well as any modern biologist. He said (1859, 1964:109):

> As new forms are continually and slowly being produced [by evolution], unless we believe that the number of specific forms goes on perpetually and almost indefinitely increasing, numbers [of species] inevitably must become extinct.

The process of competition and selection among species probably has two different, almost opposite effects. Limitation of number of species in small areas and unfavorable climates is the more obvious effect and the best evidence of the existence of the process. But the process of competition and selection among species also contributes to the evolution of better species, and probably does so most effectively in large, climatically favorable areas where species are most numerous.

Area and dispersal. Evidence of a relation between area and the effectiveness of evolution is found in the observed tendency of plants and animals to disperse from large to small areas more than the reverse. Darwin, with his usual prescience, detected a general tendency for dispersal to run from north to south on the continents, and from continents to islands, and he connected this tendency with area and with evolution. He said (p. 379):

I suspect that this preponderant migration from north to south is due to the greater extent of land in the north, and to the northern forms having existed in their own homes in greater numbers, and having consequently been advanced through natural selection and competition to a higher stage of perfection or dominating power, than the southern forms.

And he refers (p. 380) to

the more dominant forms, generated in the larger areas and more efficient workshops of the north.

This striking and appropriate phrase has been too little noticed by later evolutionists and zoogeographers. Matthew (1915, 1939) too detected a general north-to-south direction of dispersal on the continents, although he did not correlate it primarily with area (see below). Simpson (1947; 1940b), in two important papers on exchanges between mammal faunas shown by the fossil record, found more movement from Eurasia to North America than the reverse in every epoch in which evidence was adequate, and found also much more net movement from North to South America than the reverse after these two continents became connected, 1 or 2 million years ago. The exchange between Eurasia and South America (across North America) has been even more one-sided. Many originally Eurasian groups of mammals have invaded South America, but no South American mammal has reached Eurasia. Further evidence of direction comes from study of island faunas. Everywhere, most movements of plants and animals seem to have been from continents to islands. Island forms probably do sometimes re-invade continents, but they do so relatively rarely. This is, of course, only a summary of the evidence, which I have given in more detail elsewhere (1957; 1959a).

Darwin (p. 382) compared dispersing plants and animals to

the living waters . . . [that] have flowed with greater force from the north so as to have freely inundated the south.

I (1959b:314) have rephrased and extended this analogy, saying:

We know . . . that Darwin's "living waters," the invisible rivers that living things make as they evolve and flow over the earth, are not random but have a central pattern . . . This river analogy is too simple. The invisible rivers that evolving, moving animals make do not flow directly but are like tidal rivers that flow back and forth (but still have a net direction) and are so full of complex cross currents and whirl-pools that the main flow is hard to see. But underneath the details there is the main pattern that Darwin saw and that we see more clearly now: evolution of successive dominant groups in the largest favorable areas and movement (spreading) to smaller areas, from the great continents to smaller continents, and from all the continents to islands.

(I should add that, although some zoogeographers are convinced they see this pattern, others still doubt it.)

Climate and dispersal. A relation between climate and dispersal is harder to detect, and more disputed. I do not think Darwin considered it, although his idea that evolution should be most effective where there are most plants and animals implies an advantage in favorable climates as well as in large areas. Matthew (1915, 1939) did consider it, in his classic paper on climate and evolution. This was a deservedly influential contribution to zoo-geography. Nevertheless, Matthew's idea that the most successful plants and animals evolve in inhospitable climates (this *was* Matthew's idea—that the demands of "unfavorable" climate cause effective evolution) seems to me to reverse the probabilities. Matthew assumed, without putting it into words, that the challenge of unfavorable climate is greater than the challenge of (competition among) large numbers of plants and animals in favorable climates. But the reverse seems more likely: selection in unfavorable climates presumably makes climate-tolerant plants and animals, while selection in competition with other organisms presumably makes generally superior organisms, and should do so most effectively in favorable climates where organisms are most numerous. (Dobzhansky (1950) has suggested something like this, in a paper on evolution in the tropics.) Apparent histories of recently dominant groups of animals, especially mammals, and existing patterns of distribution of whole faunas (Darlington 1959a:499–502) indicate that dominant animals do most often evolve in favorable, usually tropical, climates and that the dominant groups then disperse (spread) into less favorable climates. However, perhaps the best reason for accepting this hypothesis of the relation of climate to dispersal is not the direct evidence, which is difficult to interpret, but the fact that area and climate have comparable effects on numbers of species and may therefore reasonably be supposed to affect evolution and dispersal in comparable ways.

Evolution in relation to area, climate, number of species, and dispersal. Number of species is visibly correlated with area and with climate, and dispersal goes more from large to small areas and probably also more from favorable to unfavorable climates than the reverse. This suggests that effectiveness of evolution is correlated with number of species and thus with area and climate. I have discussed this in more detail elsewhere (most recently in 1959a).

This hypothesis concerns the effectiveness of evolution, not primarily the rate of it. New species may evolve most rapidly on islands or in other small peripheral areas, but rapid evolution of small populations in such places may involve gene loss and genetic drift and is not likely to be effective in producing better plants and animals. Another possibility is that evolution may be most rapid in the tropics because of (hypothetically) higher mutation rates there and more rapid succession of generations, at least of cold-blooded animals. This may be the case, but I do not think there is much evidence of it, and it need not be so. Evolutionary rates are now thought to be under some genetic control, by selection for favorable rates of mutation or for homeostatic mechanisms (Mayr 1963:198, 288). It is therefore likely that rates of evolution everywhere tend to be brought toward a favorable mean. This is suggested also by the fact that rates of evolution in nature seem usually to be much slower than theoretically possible, slower than in the laboratory or under selection by man, as if too-rapid evolution is disadvantageous and is selected against.

However, effectiveness, not rate, of evolution is under discussion now. An analogy may help make the difference clear. Of two houses built at the same rate, one may be well built and the other badly built, the latter perhaps with defective or missing parts. And of two species that evolve at the same rate, one may evolve a good, well-integrated gene pool fitting it for further evolution and adaptation to many situations, and the other may evolve a smaller or poorer gene pool, with important genes defective or missing. The latter result probably occurs most often in small populations in small areas, and this is an indirect way of saying that evolution is more effective in larger areas. [But this is a complicated matter.

See Mayr (1963:206–207, 538).] This statement concerns evolution of single populations. The point now is that selection occurs not only among individuals in single populations but also among species and is probably most effective where species are most numerous, in large and favorable areas. Effective evolution in large areas and favorable climates makes for what I have elsewhere (1948:109; 1957: 565ff; 1959a:488) called general adaptation, which can lead to evolution of new dominant groups, as opposed to special adaptation to local environments.

Evolution in relation to size. It may be noted parenthetically, as part of the present hypothesis, that number of species is correlated also with body size, at least among animals. More species of small animals than of large ones occur together in most places, and there is therefore more opportunity for competition and selection of species among the small forms. This should give small animals a long-term advantage in effectiveness of evolution and may account for the repeated rise of new dominant groups from small ancestors seen in the fossil record, for example among mammals. See my 1959a paper for further discussion of this idea.

Summary of the evolution-dispersal hypothesis. The hypothesis reviewed in the preceding pages can be summarized as follows. Number of species of plants and animals in different places is observed to be. correlated with area and with climate. Direction of dispersal is observed to be correlated with area and probably also with climate. From these observed correlations a further correlation is inferred between number of species and effectiveness of evolution. And these observed and inferred correlations suggest the existence of a world-wide pattern of evolution and dispersal running (as the net or statistical sum of many complex movements in many directions) from large to small areas and from favorable to unfavorable climates. This hypothesis has been presented in greater detail elsewhere (see especially Darlington 1959a). It has been reviewed here, and some of the evidence for it given, to enable readers to judge the probability of existence of the evolution-dispersal pattern and the place of the southern cold-temperate zone in it.

6. Place of the southern cold-temperate zone in the evolution-dispersal pattern

The place of the southern cold-temperate zone is, of course, at the southern edge of the apparent world-wide pattern of evolution and dispersal described in the preceding chapter. Land areas south of the tropics are small to begin with, and the land narrows southward in both southern South America and Australia–Tasmania, and climate worsens southward too. As a result, according to the hypothesis, successive dominant groups of plants and animals should evolve in larger or more favorable areas and disperse (spread) southward, with either continual replacement or accumulation of relicts in the far south. The nature of far-southern floras and faunas seems consistent with this pattern of evolution and movement.

This hypothesis accounts for accumulation of relict plants and animals in the far south as the result of a process, not just of chance. To return to the river analogy, a river may deposit small amounts of drift anywhere along its length, but drift accumulates in greater quantities at special places determined by the course of the current and the nature of the banks. Similarly, the invisible rivers that living things make as they evolve and flow over the earth may leave single relicts almost anywhere, but relicts accumulate in larger numbers at special places determined by the course of the current and the nature of the land. These places include the cold southern corners of Australia and South America. Dispersal currents flow into them and end there, like real rivers sinking into the ground and depositing driftwood at their endings.

Complexity of dispersal. Some biogeographers may accept the general hypothesis outlined in the last few pages but may still doubt that it can account for accumulation of related forms in places as widely separated as southern South America and Tasmania. It can. Actual cases demonstrate it. For example *Bembidion* (pp. 45–

47) is now entering and becoming adapted to southern cold-temperate habitats in southern South America and in Tasmania independently. In fact, at least three stocks of the genus seem to have invaded southern South America and three other invasions have apparently occurred in southern Australia. Among Trechini (pp. 42–45) both of the principal terrestrial subtribes seem to have crossed the tropics in the past, and three additional stocks of the tribe extend part or all of the way across the tropics now. Regardless of directions of movement, *Bembidion* and the Trechini at least illustrate complex multiple dispersals in groups of insects that are dominant mainly in temperate regions but that have crossed the tropics repeatedly and in both Old and New Worlds.

Oaks. I think that the dispersals of any dominant groups of animals are likely to be complex enough to allow for parallel crossings of the tropics and invasions and re-invasions of separate southern cold-temperate areas. Dispersals of plants are probably often complex too. The genus of oaks, *Quercus* (which belongs to the beech family, incidentally), is noteworthy in this connection. This genus occurs principally in the north temperate zone, and it is not specially adapted for long-range dispersal—how far are acorns likely to be carried by wind? (Perhaps farther than most readers may think, but not so far as smaller seeds or winged ones.) Nevertheless, *Quercus* ranges from temperate Eurasia southward across the tropical Malay Archipelago to the mountains of New Guinea, where it meets *Nothofagus* (Robbins 1961:124), and *Quercus* ranges also from North America southward through Central America to the northern end of the Andes in South America, where it forms oak-palm associations (Verdoorn *et al.* 1945:17–18). This obviously cold-tolerant genus of trees has crossed the most formidable tropical barriers between Asia and Australia and between North and South America and seems in good position eventually to reach the south temperate zone both in Australia and in South America. If it had done so in the past, botanists might now be arguing about the direction of its dispersal!

Place of New Zealand in multiple dispersal patterns. The complexity of dispersals into the southern cold-temperate zone is suggested by the relationships of some New Zealand animals. Some frogs that reach parts of southern South America and southern Aus-

tralia–Tasmania are related in a general way: Leptodactylidae
occur in both places, although the family is widely distributed
elsewhere in America and Australia, and the southernmost American and Australian-Tasmanian forms are not directly related.
But the one genus of frogs on New Zealand is very different, with
no relatives on either South America or Australia (Darlington
1957:27–28, 526, 572). In the carabid tribe Trechini (pp. 42–
45) the New Zealand forms belong to an otherwise northern
subtribe, with no existing South American or Australian representatives. In the genus *Bembidion* (pp. 45–47) the New Zealand
species seem to be derived from an independent stock, not now
represented in Australia. And in the tribe Broscini (pp. 39–42)
the New Zealand forms are referred to a subtribe that is otherwise best represented in the Northern Hemisphere. Two genera
of this subtribe now occur, with other broscines, in southeastern
Australia (none in South America), but if they became extinct
New Zealand broscines would have the same northern relationships that the frogs, trechines, and probably *Bembidion* have. This
pattern is repeated in still another tribe of Carabidae: Sphodrini
are known only from the north temperate zone except for the
occurrence of one endemic genus on New Zealand (Britton 1959).

When I first encountered this geographic pattern of apparent
relationship of some New Zealand animals to northern rather
than to Australian or South American forms, I thought it must
be exceptional or due to faulty classifications, but it is repeated
in too many cases to be explained in this way. Some persons
think that the relationships of various New Zealand plants as
well as animals must be explained by a land connection from the
north, not by way of Australia, but directly through New Guinea
and New Caledonia to New Zealand. However, this seems to me
to be an unsatisfactory explanation too. New Caledonia and New
Zealand do not have the animals they should have if the supposed land connection really existed (see pp. 103–104). I think
the true explanation of the northern relationships of some New
Zealand plants and animals is that dispersals into the southern
cold-temperate zone have been much more complex than is usually realized, with much parallel invasion, re-invasion, and extinction, including extinction in Australia of the ancestors of
many New Zealand plants and animals, and that this complex
process has produced the complex geographic relationships of the

New Zealand flora and fauna. Incidentally, I think that biogeographers, including myself, often greatly underestimate the multiplicity and complexity of plant and animal dispersal everywhere.

Relationships among the biotas of different southern cold-temperate areas, if due to parallel invasion by related forms, must be increased by climatic selection of the invaders. The latter can be derived only from cold-tolerant groups, and representatives of the same groups may be selected in geographically separate but climatically similar areas. Among Carabidae, for example, *Tachys* far outnumbers *Bembidion* in the warmer parts of both South America and Australia, but *Bembidion* is the more cold-tolerant genus (as many details of its distribution show), and it rather than *Tachys* is invading southern cold-temperate habitats in both southern South America and Tasmania (pp. 46–47).

Vertebrates versus plants and invertebrates in the southern cold-temperate zone. Far-southern terrestrial vertebrates do not show the same pattern of geographic relationships that, for example, southern beeches and migadopine Carabidae do. South America and Australia share some characteristic groups of vertebrates including marsupials (see pages 79–81), chelyid turtles, and leptodactylid and hylid frogs, but these animals do not occur on New Zealand (except as introduced), and neither they nor any other terrestrial vertebrates show special relationships between far-southern forms of South America and Australia–Tasmania. This difference in pattern of distribution between plants and invertebrates on one hand and vertebrates on the other is fundamental in southern biogeography. It has been noted on page 1 and has been discussed at greater length elsewhere (Darlington 1960a: 663–665). Galaxiids and some other small, salt-tolerant fishes that occur around the world in the far south are not so much exceptions to the rule as special cases (see p. 38).

One suggested explanation of the difference in distribution patterns is that southern plants and invertebrates are older than existing vertebrates, that they dispersed over old land connections, and that the connections were broken before the vertebrates dispersed. Another suggested explanation is that plants and invertebrates have greater powers of dispersal and have crossed water gaps in the far south that terrestrial vertebrates could not cross. One or both of these suggested explanations may be true in special cases. For example, the fact that terrestrial vertebrates have great difficulty crossing salt water explains why so few of them have reached New Zealand, and why the few that have done so tend to persist as relicts, beyond the reach of most competitors. But New Zealand is a special case. The general difference in pattern of distribution of vertebrates as compared with invertebrates and plants in the far south can be explained neither by relative ages nor by relative dispersal powers. The trouble with these explanations is that they answer the wrong question. The question they answer or might answer is: why are there no direct relationships between special groups of terrestrial vertebrates in different southern cold-temperate areas? But the real question, which these explanations do not answer, is: why are no special groups of terrestrial vertebrates differentiated in the southern cold-temperate areas of either southern South America or Australia–Tasmania? For the fact is that no such groups exist in either place.

In South America, many distinct groups of plants and invertebrates are confined to limited areas in the far south, and many of these groups are not directly related to and not derivable from anything existing anywhere else in South America. Southern beeches, peloridiid bugs, and migadopine Carabidae are examples selected for discussion here (Chapter 3). In contrast, frogs do not occur at all in southernmost habitats in South America, and the frogs that do occur toward the south, including the endemic genera in the northern end of the Chilean forest, belong to and are presumably derived from families that are widely distributed in other parts of South America. Reptiles are even less well represented and less differentiated in southernmost South America. Various mammals, including marsupials, are known (from fossils) to have existed in South America at least as far south as Argentina since the early Tertiary, but no old, well-differentiated groups of them exist now in the extreme south. Marsupials do not reach the southern tip of the continent at all, and the endemic genera of them in the northern end of the Chilean forest belong to families that are widely distributed elsewhere in South America. A few mammals do reach even the southernmost forest and moorland, but they are not much differentiated there, and *all* of them belong to families that reached South America very recently from North America and must therefore be latecomers in the far south. These few mammals in inhospitable environments at the southern tip of South America are otters, a fox, a camel (the Guanaco), and three species of cricetid rodents. The only "old" South American terrestrial mammal that extends south of the Straits of Magellan is the Tucu Tucu, a caviomorph rodent, which reaches only the opener, warmer, northeastern part of Tierra del Fuego (Osgood 1943:121, map), and its relatives (other caviomorphs) are very widely distributed in South America. Various land birds reach extreme southern South America, but most of them are not much differentiated there, and all of them are derived from widely distributed South American groups. There simply is no special, old, independent fauna of terrestrial vertebrates at the southern tip of South America, but just a modest accumulation of more or less cold-tolerant, not much differentiated, more or less recently derived representatives of groups that are widely distributed elsewhere on the continent.

The situation is the same in Tasmania. Moderate numbers of frogs, lizards, snakes, land mammals, and land birds occur in Tasmania (Lord and Scott 1924; Darlington 1957:497–498), but they are all either actual Australian species, or related to Australian species, or species that occurred in Australia once, and they all belong to widely distributed Australian families. In Australia–Tasmania, as in southern South America, there simply is no special, old, independent fauna of terrestrial vertebrates in the far south.

It seems to me that this is the essential fact in southern biogeography that needs explaining: while many plants and invertebrates in southern cold-temperate areas in South America and Australia–Tasmania are wholly independent, neither directly related to nor derivable from anything now existing anywhere else on the adjacent continents, there are no such independent groups of southern cold-temperate terrestrial vertebrates. The preoccupation of many southern biogeographers with special situations on New Zealand may have diverted them from the importance of this general fact. It cannot be explained by differences in geologic age. Mammals, frogs, and other terrestrial vertebrates have surely been in South America long enough to evolve special southern cold-temperate groups, if they were going to do so. Nor can it be explained by differences in general dispersal power or power of crossing special barriers. The southern cold-temperate extremities of South America and Australia do not seem to have been isolated by major water barriers (see Chapters 8 and 9). Moreover, no correlation seems to exist between dispersal power and evolution of special groups in the far south. The special southern groups of invertebrates include both flightless animals, with presumably relatively low powers of dispersal, and winged insects, of which the ancestors (of the bizonal groups) have been able to disperse from one end of the world to the other; and the vertebrates that have *not* evolved special groups in the far south include both flightless forms (mammals, frogs, and so forth) and flying birds. The explanation of the different patterns of distribution of vertebrates as compared with invertebrates and plants in the southern cold-temperate zone apparently must be looked for in differences in patterns (not powers) of dispersal and in the different relations of different organisms to climate. The evolution-dispersal hypothesis outlined in preceding pages seems to give a satisfactory basis for such an explanation.

The hypothesis is that, as a result of a world-wide pattern of evolution and dispersal correlated with area and climate, plants and animals from larger or warmer areas continually make multiple invasions and re-invasions of southern cold-temperate areas, *with either continual replacement or accumulation of relicts in the far south* (p. 62). The explanation of different southern distribution patterns seems to lie in the italicized words. Climatic selection allows only cold-tolerant (but not necessarily primarily cold-adapted) organisms to enter the southern cold-temperate zone. But the histories of different organisms after they enter it differ. Different major groups of animals seem in fact to react to extreme southern climates in three different ways.

(1) Cold-blooded amphibians and reptiles (and also ants and many other important groups of invertebrates and of course many plants) apparently either cannot enter climatically inhospitable places such as the southernmost forest and moorland in South America or, if they do enter, are usually unsuccessful and do not persist there long. In the case of amphibians and reptiles (and probably many invertebrates too) the reason seems to be that they cannot satisfy their heat requirements in heat-deficient southernmost habitats (p. 25).

(2) Warm-blooded mammals and birds enter southern cold-temperate climates easily. They are inherently cold-tolerant and can exist in heat-deficient environments without much adaptive change. Therefore, those that live in such places do not evolve profound adaptations that would give them advantages over later

invaders, and the latter enter southern cold-temperate areas easily in their turn. The result is *continual replacement* in the far south, with little chance of accumulation of relicts. The existing faunas of terrestrial mammals and birds at the southern tip of South America are just what this process might be expected to produce.

(3) Some invertebrates, including some insects (and of course also some plants), are inherently cold-tolerant and can and do enter southern cold-temperate areas from or through warmer regions. But, as compared with mammals and birds, they are more dependent on and more responsive to climate. Those that enter special southern cold-temperate habitats tend to evolve profound adaptations there that give them advantages over all but the most dominant later invaders, and the latter, even though cold-tolerant, probably do not move from one climate to another as easily as mammals and birds do. The result is *accumulation of relicts* in the far south. This process, very complex in fact, seems adequate to produce the special, relict floras and invertebrate faunas that exist in special habitats on the cold southern tip of South America and in Tasmania.

Differences in survival time in the far south are probably increased by the different effects of limitation of area on different animals. Populations of invertebrates and plants can persist in small isolated areas longer than populations of vertebrates can, regardless of climate. Areas are limited in the extreme south, and may not be sufficient to support independent populations of mammals and birds or even of reptiles and amphibians for very long periods.

Although I have contrasted continual replacement (among mammals and birds) with accumulation of relicts (among some invertebrates and plants), the two processes are really one basic process occurring at different rates. Among mammals and birds, which in general seem to tolerate extreme climates rather than become profoundly adapted to them, which can change from one climate to another comparatively easily, and among which independent populations may not persist long in very small areas, the process of invasion, re-invasion, and replacement apparently goes on comparatively rapidly. Among invertebrates and plants, which are more intimately dependent on climate, among which change from one climate to another may require time for evolution of new adaptations even in cold-tolerant stocks, among which re-invasion may be further retarded by presence of older specially adapted stocks, and among which independent populations may persist in small areas for long periods, the process of invasion, re-invasion, and replacement apparently goes on much more slowly, with relicts often persisting for considerable times between re-invasions. The basic process may be compared, by distant analogy, to the action of an engine that runs at different speeds determined by friction. When friction is slight (among mammals and birds), the engine runs rapidly. When what might be called environmental friction is greater (among invertebrates and plants), the engine runs less rapidly but not necessarily less powerfully. It is then comparable to a real engine that exerts great power slowly because retarded by stiff grease.

Summary of the relation of evolution and dispersal to southern distribution patterns. The essence of the present hypothesis is that, underlying the complexity of evolution and dispersal of different plants and animals, there is a coherent, world-wide geographic pattern: successive dominant groups tend to evolve in large areas and favorable climates and to disperse (spread) into smaller areas and less favorable climates. The southern cold-temperate areas of South America and Australia–Tasmania are at the southern edge of the dispersal pattern, and these areas are therefore continually invaded and re-invaded by plants and animals from larger or more favorable areas. This process is related to climate in several ways. Climate, by its effect on evolution, in part determines the general direction of dispersal, from north to south in the Southern Hemisphere. Climate also permits only plants and animals that are cold-tolerant, with low heat requirements, to enter the coldest and most heat-deficient southern environments. And climate also retards dispersal of some organisms more than others. Among mammals and birds the retarding effect of climate is slight, invasions and re-invasions of southern cold-temperate areas occur rapidly, and existing faunas there are recent and little differentiated. Among invertebrates and plants the retarding effect of climate ("climatic friction") is greater, invasions and re-invasions of southern cold-temperate areas occur more slowly, and existing floras and invertebrate faunas in the far south are therefore older and more differentiated. The situations that actually exist at the southern tip of South America and in Tasmania seem to be products and evidence of this process.

This hypothesis accounts, as I think no other does, for the nature of the biotas that now exist in southern cold-temperate areas in southern South America and southern Australia–Tasmania, considered separately. Relationships between these two biotas are surely due partly (in *Bembidion,* for example) to parallel invasions by related stocks. Nevertheless, direct exchanges of plants and animals between different pieces of land in the far south have probably occurred too. How and when the exchanges may have occurred will be considered later, in appropriate parts of Chapters 15–21.

7. Record and significance of mammals; fresh-water fishes; accessibility of the continents

Now I shall turn from general hypotheses of evolution and dispersal to what is actually known of the geologic, climatic, and biotic histories of far-southern lands. The history of each principal piece of land will be traced, if possible, from the time of late Paleozoic glaciation to the present (see the Frontispiece for a geologic time scale). However, before turning to local histories, I shall summarize more broadly the evidence of mammals and fresh-water fishes. The fossil record of mammals shows in remarkable detail the continental connections that have and have not existed since the beginning of the Tertiary. And the distribution of fresh-water fishes confirms and extends what the mammals show. The mammals and fishes together prove certain things about the Tertiary histories of South America and Australia as wholes, although these animals are not much help in tracing the special histories of the cold southern tips of the continents. The fossil records of some other animals and of some plants give evidence of a general accessibility of continents before the Tertiary, and this evidence too will be summarized in the present chapter.

Mammals. Terrestrial mammals other than bats do not easily cross barriers of salt water. A few have reached not-too-remote oceanic islands, including the Galapagos, and they have probably done so by rafting across ocean gaps. However, not everyone agrees about this and the point is not at issue now, for the significance of mammals lies in the nature and history of whole faunas rather than in exceptional cases.

Ocean barriers do virtually stop dispersal of whole faunas of terrestrial mammals for long periods, and no other barriers do so. Deserts and mountain ranges sometimes retard movements or stop dispersal of parts of faunas for short times, but they do not stop dispersal of mammals entirely or come anywhere near doing it.

This is shown both by the distribution of existing mammals and by the fossil record. Every kind of country that can support life on land is occupied or passed through now by a variety of mammals. Among ungulates (hoofed mammals), for example, some species live in the arctic (Musk Ox and Reindeer) or on high mountains (mountain sheep and goats) and others live in or range into deserts (various antelopes, camels). And the fossil record shows that, whenever land connections are known to have existed, appropriate ungulates and other mammals have crossed in numbers whether the connections have been cold (the Bering bridge in the Pleistocene) or mountainous (the Central American isthmian link) or arid (the connection between southwestern Asia and Africa). The only barriers that stop, or nearly stop, dispersal of terrestrial mammals are gaps of ocean water.

A voluminous, nearly continuous fossil record of mammals has been found in South America extending from early in the Tertiary (part of the Paleocene) to the present. The record is mainly south of the tropics, in Argentina, but limited finds have been made in other parts of South America. The record as a whole shows the evolution in South America during the Tertiary of an endemic mammal fauna almost completely isolated from the rest of the world until late in the Pliocene. This fauna included not only a diversity of marsupials and edentates, many belonging to families that are now extinct, but also a fantastic variety of ungulates, wholly different from ungulates anywhere else in the world. Paleontologists have found fossils representing 5 or 6 orders, about 20 families, and more than 150 genera of special ungulates in South America, and this is probably only a fraction of the forms that actually existed there at different times during the Tertiary. Several of the orders and all of the families and genera of ungulates in South America before the Pliocene were endemic, unknown at any time outside South America. Moreover, these old ungulates are all now extinct, replaced by "modern" ungulates that have invaded South America, by way of the Central American land bridge, since the Pliocene. No man ever saw any of the old South American ungulates so far as we know, and we have no common names for them and no popular conception of them. Many biologists and probably even some biogeographers do not even know that they existed. Scott (1937) and Romer (1945) describe some of them; Simpson (1945; 1950) lists and discusses them and their history and significance; and I (1957:349–350) have summarized their history briefly.

During the 50 million years or so that peculiar ungulates (and certain other mammals) were evolving separately in South America, a quite different ungulate fauna existed in North America. In fact, the fossil record shows a complex succession of different ungulates in North America during the Tertiary. Details are unimportant now. The significant fact is that during most of the Tertiary there was no exchange whatever between the North American and South American ungulate faunas. There is no acceptable explanation of this fact except that North and South America were separated from each other by a water gap, as shown by Harrington's (1962) paleogeographic maps, and South America must have been isolated from all other habitable continents too during this period, for no exchange of ungulates occurred with any other continent.

Although the ungulates are most striking and most significant, the history of other mammals in South America is consistent with Tertiary isolation of that continent. Whatever may have been the earlier history of marsupials (see pp. 79–81), some major groups of them evolved in South America through most of the Tertiary without exchange with any other part of the world, and edentates were confined to South America during most of this time. However, representatives of three additional orders of terrestrial mammals did reach South America during the 50 million years or so between the beginning of the Tertiary and the Pliocene. They were a rodent, an ancestral monkey, and something like a raccoon. (Bats reached South America too, but they are a special case.) Mammals like these have more chance of rafting across water barriers than hoofed mammals do, and the nature and history of the South American mammal fauna as a whole leaves no reasonable doubt that these three pre-Pliocene Tertiary invaders arrived across water. These invasions occurred at a rate of about one per 10 or 15 million years, although they were probably not evenly spaced. What barrier except a water gap could so retard dispersal?

That South America was geographically isolated during most of the Tertiary is emphasized, by contrast, by what happened when an isthmian connection was finally formed with North America late in the Pliocene, perhaps only 1 or 2 million years ago. The fossil record of that time shows a variety of North American mammals appearing suddenly in South America and of South American ones in North America. The ones that spread southward across the new connection ranged from mastodons and big cats to mice and included various ungulates, and those that spread northward ranged from giant ground sloths and glyptodonts to opossums and porcupines. Simpson (1940b:158) has described this exchange dramatically, and I (1957:367–368) have summarized it.

Whether South America was connected by land with North America, or with any other continent, at the beginning of the Tertiary is doubtful. Harrington's maps suggest no connection from the Upper Jurassic to the Pliocene, and the record of mammals requires none. The Tertiary mammal fauna of South America evolved from a very limited set of ancestors, at first just marsupials, edentates, and ungulates or protungulates, the latter perhaps at first small and still with claws rather than hoofs. Insectivores and early placental carnivores, which existed then in the main part of the world, apparently did not reach South America and were presumably kept out by a barrier. This suggests that the few mammals that did reach South America at (or before) the beginning of the Tertiary may have crossed a water gap, as a rodent, an ancestral monkey, and a procyonid carnivore evidently did later in the Tertiary (see second paragraph above).

The fossil record of mammals in Australia is comparatively poor. There is no record at all in the early Tertiary. Nevertheless comparison of the existing Australian fauna with the record of mammals in other parts of the world justifies definite conclusions about the Tertiary history of the Australian Region. Fossils in Eurasia show that placental mammals have been dominant and diverse there through the whole Tertiary. Apparently, however, no early placentals reached Australia, for they have left no descendants there. And of later placentals, apparently only rodents (and bats) have reached Australia without the aid of man. The rodents that did so are all murids. The family Muridae is apparently of relatively recent, perhaps mid-Tertiary, origin, and rodents have a record of "island hopping" in other parts of the world, so the murids presumably reached Australia across mid-Tertiary or late-Tertiary water barriers that other mammals failed to cross (Simpson 1961:432–436). Moreover, placental mammals are now thought to be as old as marsupials in their origins, so the presence of marsupials but not of old placentals in Australia suggests that the marsupials originally island hopped to the continent across water barriers that blocked early placentals. And there are hints that, even before this, monotremes may have been isolated in Australia. No fossil or living monotreme has ever been found outside the Australian Region. The two existing monotreme types, the Platypus and the echidnas, are more specialized than any marsupials, and it may be guessed that they are the last relics of a monotreme fauna that radiated in isolation in Australia in the Mesozoic and that has been replaced, except for the uniquely specialized relics, by the later radiation of Australian marsupials. I have discussed all this in somewhat different words elsewhere (1957:362–363). The conclusion, based primarily on the fossil record of mammals in Eurasia in comparison with the existing Australian mammal fauna, is that the Australian Region, including New Guinea, has been isolated through the whole Tertiary and perhaps much longer.

In short, the evidence of mammals shows that both South America (until the late Pliocene) and Australia have been cut off from the northern continents, and also from each other, by ocean gaps at least since very early in the Tertiary. This is the minimum amount of isolation that the evidence actually requires, and all biogeographers should accept it as fact. The geologic evidence of the submergence of northwestern South America suggests that the isolation of South from North America actually began in the late Jurassic. Biogeographers should, I think, take this as probable but need not be bound by it. The geologic history of the East Indies (Umbgrove 1949) has been more complex and many details are still unknown. So far as I can judge, there is no clear evidence as to whether or not Australia and New Guinea were connected by land with Asia in the Mesozoic (see especially Umbgrove, pp. 36ff), and biogeographers are therefore free to draw what conclusions they can from their own evidence. It seems to me that the Australian mammal fauna, which includes what seem to be extremely specialized remnants of a pre-Tertiary monotreme fauna confined to Australia, hints at isolation of that continent at least since some time in the Mesozoic.

Marsupials. Because they are extraordinary animals, conspicuous and well known, and because their distribution is remarkable, marsupials have long been emphasized by zoogeographers. Their

occurrence chiefly in South America and the Australian Region
has frequently been taken as evidence of southern land bridges or
continental drift. However, what is now known about their phylogeny and fossil history seems to show that they really reached
South America and Australia independently, by parallel invasion
from the north (Fig. 15), as *Bembidion* has done (pp. 45–47). After
their initial dispersal, marsupials have apparently radiated independently on the two southern continents without direct exchange
between the two but with striking convergence especially of carnivorous forms, Borhyaenidae (now extinct) in South America and
Dasyuridae (living) in Australia. I cannot here give the arguments
for and against this hypothesis but can only refer to Simpson's
(1945:165, 170–172) authoritative discussion of the evolution and
relationships of marsupials and my (1957:344–345, 362) summary
of their probable geographic history. Marsupials have always (so
far as the record goes) been associated in South America with
edentates and peculiar ungulates, which are unknown in Australia. This is additional evidence (additional to the evidence of the
fossil record and apparent phylogeny of marsupials themselves)
against a direct dispersal between South America and Australia
within the time of the animals concerned.

Whatever their history may have been, existing marsupials do

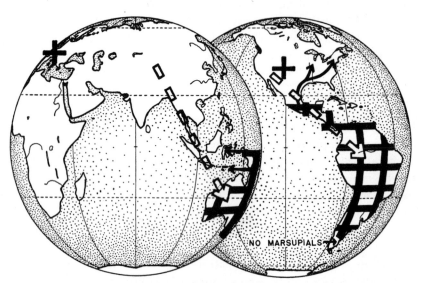

Fig. 15. Present distribution of marsupials (*solid bars and small arrows*) and their
hypothetical early dispersal (*large broken arrows*) (from Darlington 1957, Fig. 39,
copyright John Wiley & Sons).

not have a pattern of distribution like that of the far-southern groups of plants and insects discussed in Chapter 3. No special groups of marsupials are confined to southern cold-temperate areas, or at least no groups have evolved there independent of adjacent continental faunas. Marsupials do not occur at all on the southern tip of South America, and the Tasmanian ones are all either Australian species or recently derived from Australia. The "Tasmanian Wolf" and "Tasmanian Devil" are fossil in Australia and derived from a family that is widely distributed in Australia and New Guinea. Moreover, marsupials do not occur on New Zealand, or rather are not native there. (The common Australian possum was introduced into New Zealand and has proved to be a pest (Wodzicki 1950).) These discordant details are usually ignored by botanists and entomologists who wish to fit marsupials into the geographic pattern of far-southern plants and insects.

In summary: what is now known of the probable phylogeny, fossil record, and present distribution of marsupials suggests (1) that they reached South America and Australia separately, from the north; (2) that no direct land connection has existed between South America and Australia since these animals dispersed; and (3) that, although marsupials have been on the southern continents for a long time, they have not evolved independent stocks specially adapted and confined to southern cold-temperate areas and do not have a pattern of distribution like that of the southern beeches, peloridiid bugs, or migadopine Carabidae.

Fresh-water fishes. Some fishes, including some large families that are dominant in fresh water, cannot tolerate salt water or at least do not ordinarily cross salt-water barriers. A narrow barrier of salt water can stop dispersal of whole faunas of true fresh-water fishes for long periods. This has happened at Wallace's Line: many tropical Asiatic fishes reach Borneo, but no strictly fresh-water fish has got across Makassar Strait to Celebes without the probable aid of man. Nevertheless, even the greatest fresh-water fish families sometimes include exceptional, salt-tolerant species that can cross narrow barriers of salt water, so that during very long periods even ocean gaps may not always completely stop the spread of fresh-water families. For further discussion of the very complex relationships of fishes to fresh and salt water see Myers (1938) and Darlington (1957:41–47).

South America and Australia are remarkable for the fresh-water

fishes they have and do not have. The faunas of these two continents have nothing in common so far as primary groups of fresh-
water fishes are concerned, although they do share galaxiids and
some other salt-tolerant groups in the far south (see p. 38). South
America has an enormous fauna of true fresh-water fishes, which,
however, have evolved from very few ancestors (Darlington
1957:69–72, 91–99). Australia and New Guinea have no undoubted
"true" fresh-water fishes except the Australian lungfish, a relict
which is now confined to a small area in subtropical southern
Queensland and is not directly related to the South American
lungfish. An osteoglossid in tropical Australia and New Guinea
may or may not be a strictly fresh-water fish; in any case its only
close relative is Oriental, not South American. Regardless of how
the fishes have dispersed, the fewness of ancestors and the extent
of radiation of some groups on South America suggest long isolation of that continent. And the absence of any comparable fresh-
water-fish fauna in Australia and New Guinea suggests that the
whole Australian Region has been isolated by barriers of salt water
either more effectively or for a still longer time.

*Summary of isolation of South America and Australia; unity of South Amer-
ica.* The fossil record and the present distribution of mammals
show conclusively that South America (until the Pliocene) and
Australia have been isolated from other habitable continents and
from each other at least since very early in the Tertiary. There
are indications that the two continents in question may have been
isolated for a much longer time, although this is not certain. The
distribution of fresh-water fishes as well as of mammals is consistent with longer isolation, and so is the geologic evidence.

   Mammals and fishes show one other thing about the history of
South America: the continent has been essentially one piece of
land for a long time. The occurrence of mammals, some apparently derived from North America, as far south as Argentina early
in the Tertiary suggests that South America was continuous land
then. The nature of the fish fauna indicates the same thing. Only
one main fauna of true fresh-water fishes exists in South America.
It centers in the Amazon and parts of it have spread from there in
all possible directions (Eigenmann 1909:371; Darlington 1957:72).
And unity of South America is indicated by geologic evidence
too; Harrington's (1962) maps show no division of the continent

into major parts since the Paleozoic (but see p. 87). Von Ihering (1900 and other papers) and other persons, including Eigenmann, have thought that South America was formerly divided by an ocean gap into two wholly separate parts, on which separate faunas of fresh-water fishes and other animals evolved independently, but this idea was evidently mistaken. The evidence against it, summarized above, is very strong now. The striking differences that do exist between the faunas of the tropical and south-temperate parts of South America, and which so impressed von Ihering and Eigenmann, are evidently due to climatic differences rather than to former division of the continent by a water gap.

Accessibility of the continents. In spite of the fact that both South America and Australia are known to have been isolated for long periods by at least narrow ocean gaps, the fossil record of terrestrial plants and animals shows that all the habitable continents have been reasonably accessible to land life at least from time to time since the late Paleozoic. Ancient plants, including early conifers, spread over the whole world before the end of the Paleozoic (Kräusel 1961, with references). Terrestrial reptiles, including dinosaurs, reached all continents (apparently excepting Antarctica) during the Mesozoic. The known distributions of reptile groups in the Upper Permian, Lower Triassic, and Middle and Upper Triassic are diagrammatically mapped in Fig. 16. These maps emphasize two things: the incompleteness of the geologic record and the closeness of relationships between the reptile faunas of the Northern and Southern Hemispheres. And angiosperms spread over the whole world in the Cretaceous if not earlier, and their dispersal included many exchanges between northern and southern continents, including north-south dispersals between Asia and Australia.

Barriers have surely existed in some places at some times, including the Tethys Sea across southern Eurasia in the Mesozoic and the water gaps that have isolated both South America and Australia more recently, but the barriers have been incomplete, or temporary, or so narrow that some terrestrial plants and animals have crossed them. Accessibility of the continents in spite of the existence of moderate or temporary barriers has been the general rule. Specifically, the world has not been divided into northern and southern supercontinents so widely separated that dispersal has been stopped for any important span of time. When, therefore,

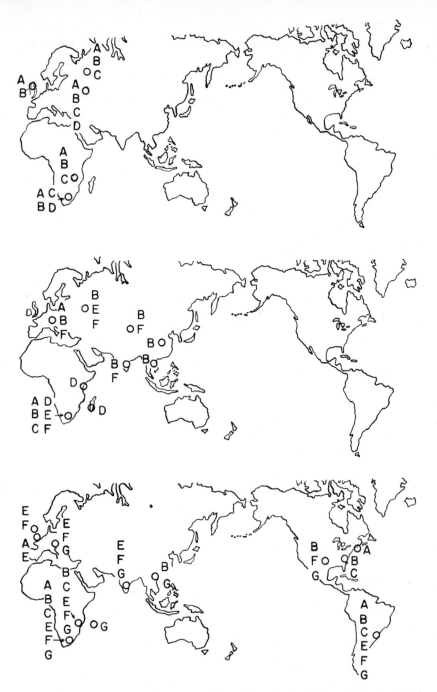

Fɪɢ. 16. Distribution of reptiles in the Upper Permian (*top*), Lower Triassic (*middle*), and Middle and Upper Triassic (*bottom*): *A*, stem reptiles, cotylosaurs; *B*, plant-eating mammal-like reptiles; *C*, carnivorous mammal-like reptiles; *D*, Eosuchia; *E*, rhynchocephalians; *F*, thecodonts; *G*, dinosaurs. (After S. H. Haughton, 1953. From *History of the earth: An introduction to historical geology* by Bernhard Kummel. San Francisco: W. H. Freeman and Company, 1961.)

some ancient plants are found to have been distributed in separate northern and southern zones, as early coal floras and conifers were, the zonation presumably reflects differentiation of climates rather than separation of the zones by impassible barriers.

8. History of southern South America

This chapter on southern South America (see map, Fig. 2) is the first of six chapters summarizing what is known or can reasonably be deduced about the histories of the principal pieces of land in the far south. In each case, ideally, I should like to present geologic-geographic, climatic, and biotic histories separately, from the late Paleozoic through the Mesozoic, Tertiary, and Pleistocene to the present. This, however, is still an unrealizable ideal. Histories are not known in such detail, and different kinds of history are too interdependent to be completely separated. For example, glaciation is involved in both geologic and climatic histories; the geographic history of land connections between continents is often best shown by the history of faunas; and climatic history is often best shown by the history of vegetation.

In tracing histories, time and knowledge are inversely correlated: in general, the longer the time since a given event, the less is known about it. The following histories are therefore most detailed and most trustworthy in the most recent geologic periods. The histories will be sketchy at best in the Paleozoic and early Mesozoic, better but still doubtful in many details even in the late Cretaceous, and still better and in most essentials trustworthy in the Tertiary.

Geologic-geographic history. The geologic history of South America as a whole, up to the Pliocene, is traced by Harrington (1962) in 46 maps. The maps show known marine, continental, and mixed or brackish-water as well as volcanic and glaciated areas for successive geologic intervals. These maps show several facts important to biogeographers.

First, the maps indicate that the northwestern corner of South America, toward what is now Central America, was covered by the sea at least from the Upper Jurassic through the Cretaceous and most of the Tertiary until the Pliocene, suggesting that South America was isolated through this whole time. The geologic record thus agrees with and extends the evidence of mammals and fresh-water fishes (Chapter 7), which requires isolation of South America in the Tertiary but only hints at it earlier.

Second, Harrington's maps suggest that South America has been essentially one piece of land from the late Paleozoic to the present. Zoogeographic evidence (Chapter 7) suggests the same thing, but covers a shorter span of time. The western edge and southern tip of South America have been submerged at times (partly in the course of complex geologic processes that preceded formation of the Andes), and a fluctuating archipelago has probably existed at the southern tip of the continent from time to time, as now. (However, although Harrington's maps show no substantial part of southern South America isolated for long, some geologists think the continent may have been divided by seaways at least for short periods (Durham 1964:357, Fig. 1).) In any case, the existing Andes were not formed until the Tertiary.

Glaciation of continental proportions occurred in southern South America in the late Carboniferous (Upper Pennsylvanian). Glacial tillites of this age are widespread (Harrington, pp. 1788–1789, Fig. 13), and found with them are fossil plants (*Glossopteris* and others) that are associated with Permo-Carboniferous glaciation on other southern continents. The ice sheets occurred in what is now the southern warm-temperate part of South America and the southern edge of the tropics. Signs of glaciation at this time have apparently not been found on the extreme southern tip of South America, although ice did occur on the Falkland Islands. The continental glaciation apparently ended about the beginning of the Permian, although Alpine-type glaciers seem to have existed locally in the region of present-day Bolivia and northwestern Argentina in the Upper Permian (Harrington, p. 1791 and Fig. 15). No further significant continental ice occurred on South America until the Pleistocene.

During the Pleistocene, the southern tip of South America, including Tierra del Fuego, was repeatedly glaciated (Flint 1957:426–429, Fig. 24-3; Auer 1960), and the glaciations were probably contemporaneous with those in the Northern Hemisphere. This glaciation centered much farther south in South America than the Permo-Carboniferous one did, and was much less extensive.

Fluctuations of sea level occurred in southern South America in the Pleistocene, probably as part of world-wide changes of sea

level correlated with continental glaciations (Auer): when water
was transferred from the sea to ice on land, the level of the sea
fell everywhere. The Straits of Magellan are less than 50 m deep
in places, according to the current map published by the American Geographical Society (1956). Presumably, therefore, Tierra
del Fuego was repeatedly connected to the mainland, and the last
connection probably ended only about 10,000 years ago.

Climatic history. In the late Carboniferous and early Permian, most
of what is now warm-temperate southern South America evidently
had a glacial climate. The main part of South America north of
the area of glaciation was evidently warmer, but how warm it was
I do not know. The extreme southern tip of the continent may
have been warmer-than-glacial too, although this is not certain.

A general amelioration of the climate of southern South America evidently occurred early in the Permian, and the whole continent (except on high mountains) has been at least warmer-than-
glacial since then, until the Pleistocene. Except for this general
fact, nothing that is useful now can be said (I think) of the climate
early in the Mesozoic, but something can be deduced about conditions later, during and after the Cretaceous.

Much of northern South America has evidently been fully tropical for a long time, long enough to permit evolution of the complexly organized tropical rain forest. However, the dominant rain-
forest trees are angiosperms, so the existing forest may be no older
than the Cretaceous, which is apparently when angiosperms dispersed. In far-southern South America, the presence of *Nothofagus*
through much of the Tertiary indicates a cooler climate, differentiated from the tropics, fairly well watered, and at least partly
rain-forested. And the hot tropical and cooler far-southern areas of
heavy rainfall have probably long been separated by a broad barrier
of drier country, as they are now.

This pattern of differentiation of wet tropical and wet south-
temperate areas separated by drier country is determined by the
basic circulation of wind and rainfall in the Southern Hemisphere,
which in turn is determined by the shape and motions of the
earth. The main climatic pattern has therefore probably existed as
long as South America has occupied its present position on the
earth, which it has done (approximately) at least since the Jurassic, according to paleomagnetic data. However, the intensity and

details of the pattern have probably changed from time to time. In general, southern South America, like most other now-temperate regions, has probably been warmer than now (but not tropical) most of the time between the late Paleozoic and Pleistocene glaciations. The arid parts of southern South America were evidently wetter than now during Pleistocene pluvial periods. And the rise of the Andes must have affected local climates complexly. The Andes eventually formed a temperate corridor across the tropical part of South America, but probably not until the Tertiary.

The southern tip of South America now has an "oceanic" climate and probably has had one as long as the land has not changed too much in position or area. This climate is cool or cold but without the extreme fluctuations of temperature that occur on north-temperate continents. The southern cold-temperate oceanic climate of southernmost South America is in some ways more like the montane climate of the equatorial Andes and of other high mountains in the tropics than like the climate of north-temperate continental areas (Troll 1960). This important fact will be considered in more detail in Chapter 14.

All these facts and inferences suggest the following climatic history for the southern tip of South America. The southernmost tip of the continent may have been warmer-than-glacial even when some other parts of southern South America were glaciated in the late Carboniferous and early Permian. Since then the tip has usually been warmer than now, with a high-rainfall, oceanic climate, at least along the western edge of the land. The far-southern area of high rainfall has probably always been isolated from the wet tropics by a barrier of relatively arid country, although the barrier has probably fluctuated. The rise of the Andes during the Tertiary formed a temperate corridor across the tropics which lessened the climatic isolation of southernmost South America. Finally the climate of the tip became cooler in the late Tertiary and glacial in the Pleistocene, and is still cold-temperate at sea level and glacial on the mountains.

Biotic history. Much of our information about the history of plants and animals on the southern tip of South America, as well as much information about present distributions there, is contained in a set of papers on the biology of the southern cold-temperate zone published by the Royal Society (Pantin 1960) following a symposium held in London in December 1959, which in turn followed a Royal Society expedition to southern Chile.

A specialized, southern-hemisphere flora of *Glossopteris* and associated plants occurred in the glaciated part of southern South America late in the Paleozoic. I do not know whether this flora reached the extreme southern tip of the continent.

Conifers occurred well south in South America perhaps before the end of the Paleozoic and surely in the Mesozoic. Araucarians especially are fossil there in the Jurassic (and later), but were widely distributed on *northern* continents at the same time (Florin 1963:178).

Direct evidence, especially the pollen record, shows that *Nothofagus* has been in southern South America through much of the Tertiary, at least since the Eocene (Couper 1960:493). In general the flora of far-southern South America was richer in the Tertiary than now, for many species of plants became extinct there during the Pleistocene (Auer 1960:507, 516). However, additional plants have probably been invading southern South America from the north from time to time, some of them recently. This is indicated by the numbers of amphitropical genera and species that now occur, often discontinuously, in north-temperate and south-temperate areas in the Americas (Constance 1963). On the other hand, many southern plants have spread far northward along the Andes. This is inferred from the vegetational resemblances between the subantarctic moorland of extreme southern South America and the equatorial high Andes (Troll 1960:531), and from the fact that the Andes are relatively recent in origin. The northward spreading has evidently been facilitated by the climatic similarity (noted above) of the southern cold-temperate and high-montane climates. These plants have not moved against the hypothetical dispersal pattern described in Chapters 5 and 6 (although some counter-movements are to be expected), for they have not invaded larger or climatically more favorable areas but have simply spread into what is virtually a new extension of the habitat they were adapted to.

Complex local changes in distributions of some plants in far-southern South America during the Pleistocene, especially of *Nothofagus,* are partly demonstrated by the pollen record (Auer) and partly inferred, for example from the fact that the plants now extend into recently glaciated areas. A fact stressed by Kuschel (1960:544) and fully confirmed by my experience, that no distinct species of animals are confined to the southern end of the Chilean forest, is consistent with a recent extension southward both of the forest and of the animals in it. Additional details of Tertiary and Pleistocene as well as of present distributions of plants in far-southern South America can be found in the Royal Society symposium referred to above, especially in contributions by Skottsberg and Godley.

Terrestrial animals have left comparatively little direct evidence of their past distributions and movements in far-southern South America. Some of the distinct groups of invertebrates that now occur in southern cold-temperate habitats have presumably had long geographic histories like those of the older plants. Existing southernmost mammals have invaded the far south much more recently, since the late Pliocene, for they all belong to families that did not reach South America until then (see p. 77). Southernmost land birds have probably reached the southern tip of the continent recently too.

The Tertiary and post-Tertiary biotic history of far-southern South America can be summarized as follows. The genus *Nothofagus* has existed there at least since the Eocene in company with many other plants, of which some survive, some have become extinct, and some (principally moorland forms) have spread northward along the Andes, while additional plants have probably been arriving continually from the north. Some invertebrates have probably had similar Tertiary histories, but southernmost mammals and birds are relatively recent, perhaps Pleistocene or post-Pleistocene invaders. All this is reasonably consistent with the hypothesis (pp. 68–70) of continual invasion and re-invasion of the southern cold-temperate zone, with survival times of the invaders varying with different organisms' different relations to the environment. But allowance must be made for the diversity and complexity of actual cases. Also, this pattern of north-south dispersals does not rule out the possibility of dispersals occurring also around the southern end of the world.

9. History of the Australian Region, especially Tasmania

For a map of the Australian Region, including Tasmania, see Fig. 3.

Geologic-geographic history. The geologic and geographic history of Australia and Tasmania should, for biogeographic purposes, include the history of the Indo-Australian (Malay) Archipelago west at least to the continental shelf of Asia, the edge of which coincides in part with Wallace's Line (Darlington 1957:462). This archipelago has had a complex history (Umbgrove 1949), which is not fully known. However, the archipelago itself has not produced geologic evidence of any complete land connection between Asia and Australia, and the distributions of mammals and fresh-water fishes (Chapter 7) show conclusively that Australia has not been connected with any other habitable continent since the beginning of the Tertiary, and perhaps not then. The geographic isolation of Australia and New Guinea by water gaps through the Tertiary, and perhaps for a much longer time, is a fact that some biogeographers have been slow to appreciate. Additional evidence of isolation is found in the fact that the flightless carabid beetles on high mountains in New Guinea have apparently all evolved *in situ* from winged ancestors (most of which apparently came from surrounding tropical lowlands) and are not related to flightless Carabidae of distant temperate regions (Darlington, unpublished). This is an indication that the existing mountains of New Guinea have not been connected by continuous ranges with mountains in either Asia or Australia (and see following paragraph).

Within the Australian Region, New Guinea has been notably unstable. The physiography of this and adjacent islands "reflects the dominating influence of young and intense orogenic processes" (David 1950, vol. 1, p. 681). Details are apparently not very well known. My impression is that the existing mountains of New Guinea are indeed rather recent in origin and that the existing mountain fauna, like the mountains themselves, is relatively recent.

I know that the carabid beetles on mountains in New Guinea seem less old (less differentiated, with fewer relicts) than those of, for example, New Caledonia or New Zealand. However, I know no way to determine the actual time of origin of the mountain-living plants and animals now on New Guinea. They have left no fossil record or at best, as in the case of *Nothofagus* (Chapter 16), the record may be incomplete.

The geologic-geographic history of Australia, necessarily over-simplified and partly hypothetical, can be followed by means of David's maps, especially the ones in volume 1 on pages 393 and 394 (Permian), 436 (Triassic), 474 (Jurassic), 511–513 (Cretaceous), and 583 and 584 (Oligocene, Miocene). Australia has been a relatively stable block of land throughout the time under consideration, from the late Paleozoic to the present, although parts of the continent were sometimes invaded by shallow seas. The latter seem to have cut the continent into two or three separate parts for a while in the Lower Cretaceous. Biogeographically, the epicontinental seas and the enormous lakes characteristic of Australian geologic history may have been less important as barriers than for their effects on climate.

Glaciation occurred on a "stupendous" scale (David, vol. 1, p. 369–371, Fig. 116) in Australia in the late Paleozoic, from the Carboniferous into the Lower Permian. Ice sheets reached into what is now the edge of the tropics on higher land and came down to the sea at least as far north as 29° S on the west and 32° S on the east side of the continent (David). Southward, ice reached the southern limit of land, including what is now Tasmania. The direction of movement of the ice is, I think, still somewhat doubtful: David's description of the direction of ice movement is at least partly wrong (see p. 182).

A report of glaciation of part of southern Australia in the Lower Cretaceous (David, pp. 503–505, Fig. 149) was apparently errone-ous (King 1961:330, quoting Parkin).

Pleistocene glaciation of Australia was very limited, apparently confined to the highest mountains of the southeastern corner of the continent and the higher parts of Tasmania (David, pp. 623ff; and see following paragraph). However, world-wide eustatic lowerings of sea level, which were caused by transfer of water from the sea to ice on land on other continents, and which connected Tierra del Fuego to South America (Chapter 8), also connected both New

Guinea and Tasmania to the Australian mainland. This can be
deduced from the shallowness of the water that now separates these
islands from Australia. And in each case the connection is proved
by the identity of many species of mammals on opposite sides of
the present water gap. The connections were probably made and
broken several times, during successive Pleistocene glaciations and
interglacial periods. The last connections were presumably broken
about 10,000 years ago.

A special review of Tasmanian geologic history has been published as Part 2 of Volume 9 of the *Journal of the Geological Society of
Australia* (Spry and Banks 1962). Articles in this review provide the
following items of present interest. Although the corner of Australia
that is now Tasmania may have been entirely submerged at times
earlier in the Paleozoic (Solomon 1962:311), it seems to have been
at least partly land from the later Paleozoic to the present. Widespread glaciation occurred in the late Carboniferous or Permian
and was followed by deposition of thick fresh-water and marine sediments, some of which contain *Glossopteris* and other Permian fossil
plants, and oil shales and coal occur in the Permian of Tasmania
(Banks 1962). Succeeding Triassic rocks are entirely fresh-water and
include coal measures (Solomon, p. 335). The record of the later
Mesozoic is relatively poor. The earth movements that have given
Tasmania its present shape and surface occurred mainly in the Tertiary. At the beginning of the Tertiary, Tasmania was still part of
Australia. It became separated by formation of Bass Strait perhaps
as early as the Oligocene (Gill 1962:248–249), although I think the
times when Tasmania has and has not been connected to the mainland before the Pleistocene are somewhat uncertain. Pleistocene
glaciation of Tasmania was limited to the central plateau and
scattered mountains, and covered only a small part of the island's
area (Gill, p. 246, map).

Climatic history. Climatically, southern Australia was glacial in the
late Carboniferous and early Permian, warmer thereafter (perhaps
warmer than now for long periods but probably not tropical),
then cooler again late in the Tertiary and locally glacial in the
Pleistocene. Coal was laid down in southern Australia at various
times in the Permian, Triassic, Jurassic, and Cretaceous (David),
but this may indicate no more than a cool-temperate climate
with sufficient rainfall. Some of the Permian coal beds lie be-

tween or immediately above glacial deposits. This simple summary of late Paleozoic and Mesozoic climatic history presumably covers a diversity of minor fluctuations and local variations.

The climatic history of the Australian Region during and since the Tertiary, and perhaps before, has probably paralleled that of South America, for the two land areas have the same general relation to the ruling pattern of wind and rainfall in the Southern Hemisphere. New Guinea and the northern edge of Australia are and probably long have been tropical, with extensive or (in northern Australia) local areas of heavy rain and tropical rain forest. Tasmania and the southern edge of Australia, where *Nothofagus* has existed since the late Cretaceous, have evidently been cooler and at least partly wet, with extensive or local tracts of south-temperate rain forest. And between the tropical and south-temperate areas of heavier rainfall there is and probably long has been a broad barrier of more or less drier country (but see following paragraphs). However, while the arid or semiarid barrier in South America is now, I think, complete from coast to coast across the Andes, the barrier in Australia is not complete. The whole eastern edge of Australia receives more rain than the interior (but not as much as parts of tropical Queensland or of Tasmania) and is variably forested, and this is likely to have been true (with variations) as long as the relation of land to ocean has been the same as now. There is therefore a corridor along the eastern edge of Australia open to some plants and animals that cannot tolerate acute aridity. The aquatic Platypus, for example, occurs along this corridor from tropical North Queensland to south-temperate Tasmania. Flightless forest-living Carabidae, too, extend along this corridor from north to south, the tropical and south-temperate groups meeting and overlapping complexly (Darlington 1961a).

The preceding paragraph must be qualified. There is evidence, summarized by Gentilli (1961), that warm, rainy climate extended far southward in Australia in the Eocene and early Oligocene. Trees that are now confined to tropical or subtropical Queensland occurred in southern Australia and Tasmania, and heavy forests grew in what is now arid country near Lake Eyre and in southwestern Queensland. Because of this it has become rather fashionable (as Gentilli says) to contrast a "tropical" Tertiary with a "glacial" Pleistocene in Australia. However, Gentilli (p. 469) thinks that, while it is reasonable to conclude that the Eocene and

Oligocene climate of southern Australia was wet and subtropical (or warm-temperate—the terms are not exact), there is no basis for the belief that conditions were actually tropical. Burbidge (1960: 160) seems to have reached the same conclusion. And the diversity of *Nothofagus* in southern Australia during the Tertiary indicates a less-than-tropical climate.

The hypothesis of tropical Tertiary Australia, based on doubtful evidence to begin with (see above), has consequences that raise further doubts. The genera *Eucalyptus* and *Acacia* are now very numerous in species and very widely distributed as dominant trees in the drier, opener forests and dry scrubs of Australia, but they are poorly represented in rain forest. Because he thinks that Australia was entirely tropical and humid in the Tertiary, Gill (1961:338) concludes that the *Eucalyptus* and *Acacia* woodlands must be "essentially Quaternary." But I cannot believe this. I am thoroughly familiar with the *Eucalyptus* woodlands of eastern Australia—I have traveled and lived in them in a small truck for many months— and I know that *Eucalyptus* carries an enormous fauna of special insects. Many of them feed on *Eucalyptus* nectar. Many others eat the leaves or bore in the wood. And a whole special fauna of predaceous insects lives on *Eucalyptus* trunks, hiding by day under flakes and slabs of loose bark on living trees. These *Eucalyptus*-trunk predators include about 200 species of Carabidae belonging to many genera derived independently from 6 principal tribes; some genera and even one subfamily (Pseudomorphinae) occur only or mainly in this habitat; and striking convergence of color patterns has occurred among species derived from different ancestors. The existence of this extraordinary insect fauna implies that *Eucalyptus* woodland has been a major habitat in Australia for a long time, not just during the Quaternary. The desert and semidesert habitat too is evidently old, with a very distinct fauna which includes the Marsupial Mole (an endemic family), other desert-living mammals that are generically distinct, and many special groups of carabid and tenebrionid beetles and ground-living weevils.

The apparently conflicting evidence can, I think, be reconciled in one consistent hypothesis. The hypothesis is that the whole width of southern Australia was warmer and wetter in the Tertiary than it is now, the heavier rainfall perhaps being due partly to northward incursion of the sea (Gentilli, p. 469); that on the eastern side of Australia a strip corresponding to but wider than the existing "corridor" had heavy rainfall too; but that at the same time a large part of interior and western Australia was comparatively dry. In other words, according to this hypothesis, the present broad pattern of climate over Australia was the same in the Tertiary as now, as would be expected, but the southern part of the continent was warmer and the area of rainfall in the south and east was wider and wetter than now (Fig. 17).

This discussion cannot do justice to the complexity of the climatic history of Australia. Climatic oscillations probably reached a maximum during the Pleistocene, although the successions of wet and dry climatic phases have not been fully worked out (Burbidge 1960:183). Keast (1961:424ff), to whom, incidentally, I am indebted for recent comments and references, summarizes what is known and inferred. During the Pleistocene, glaciation was negligible on the continent (and not extensive even on Tasmania; see page 94), but rainfall fluctuated and was rather high in some areas at some times. Authorities do not agree whether all Australia was wet at these times or whether there were shifting zones of rain and aridity, but it seems to me, for reasons just given, that large parts of the continent must have been relatively dry even when rainfall was heaviest.

I suspect that both the amplitude and the effect of Pleistocene fluctuations of rainfall have been overestimated. Again there is evidence from Carabidae. At least two eastern Australian genera of flightless forest-living carabids, which are good indicators of continuity of habitats, are represented in the isolated forests of southwestern Australia. One genus, *Trichosternus,* now occurs mainly in the warmer part of eastern Australia, from tropical Queensland to central New South Wales, with one very distinct species in southwestern Australia (Darlington 1961b). And the other, *Notonomus,* occurs in eastern Australian forests from tropical Queensland to Tasmania, with one very distinct species isolated in the southwest (Darlington 1961c). The point is that the southwestern representatives of these two genera are very distinct species. They are apparently Tertiary rather than Pleistocene relicts (cf. p. 12), and no less-differentiated representatives of significant eastern Australian carabid genera occur in the southwestern forests. Southern Australia was apparently not wet enough during the Pleistocene to allow formation of a continuous strip of even moderately wet forest across the continent, although grassland probably was continuous at times.

The present climate of Tasmania is "oceanic," comparable to

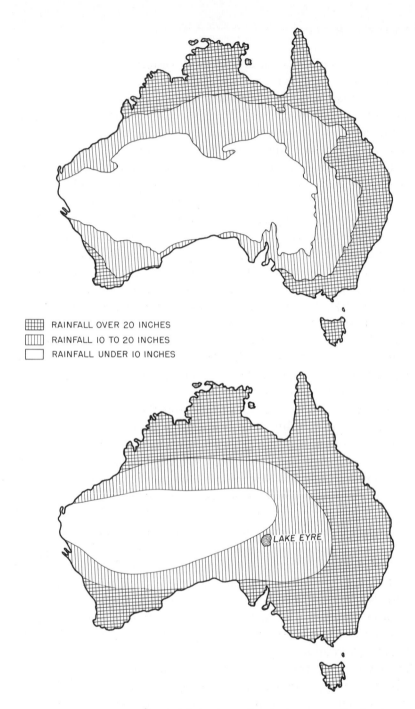

RAINFALL OVER 20 INCHES
RAINFALL 10 TO 20 INCHES
RAINFALL UNDER 10 INCHES

LAKE EYRE

FIG. 17. Distribution of rainfall in Australia: (*top*) present distribution (from CSIRO 1950, Fig. 10, redrawn and simplified); (*bottom*) hypothetical distribution in the past.

the climate of the South Island of New Zealand (*Australian Ency-clopaedia*), and the nearness of the ocean may have buffered climatic changes in Tasmania in the past.

Biotic history. The biotic history of southern Australia and Tasmania has been complex too, but can be only briefly summarized here. The *Glossopteris* flora occurred in glaciated areas in late Carboniferous and Permian times. Deposition of coal then and during the Mesozoic (p. 94) indicates persistence of fairly heavy vegetation on Tasmania, perhaps some sort of south-temperate rain forest, composed at first of nonangiosperms. During at least part of the Tertiary *Nothofagus* apparently occurred across the whole of southern Australia (Burbidge, p. 162), and then or later (Pleistocene?) it evidently extended also northward along the eastern edge of Australia at least to southern Queensland and perhaps farther. Fossil floras show a variety of other trees too, in southern Australia in the Tertiary, some south-temperate and some more or less tropical in present relationships (Burbidge, p. 160). They suggest, to me, not necessarily a continuous stretch of mixed forest but a patchwork like the one I have seen existing now in parts of Tasmania, with *Nothofagus* forming heavy stands in suitable places and other types of forest interspersed.

So far as I know, no fossil record has been found of the insects and other animals that lived in the Tertiary forests of southern Australia, but scattered relicts suggest that the insects, like the trees, were a mixture of what are now tropical or warm-temperate and what are now cool-temperate groups. Among Carabidae, for example, the *Trichosternus* that is relict in southwestern Australia (see p. 97) belongs to what is now otherwise a tropical and warm-temperate genus, while two relict genera of Migadopini known only from very restricted areas in eastern Australia (Fig. 9) belong to an "antarctic" tribe. These genera are *Nebriosoma,* found near Kiama (34°–35° S), and *Decogmus,* found near Comboyne (31°–32° S). The existence of these relicts indicates that *Trichosternus* and Migadopini occurred together in southern Australian forests in the Tertiary.

Two things have happened to the rich, diversified forests, and to the probably correspondingly rich and diversified faunas of insects and other animals associated with the forests, since their maximum development across southern Australia perhaps early

in the Tertiary. First, some of the constituent groups have become extinct at least in Australia, and these include even one of the three main groups of *Nothofagus* (see below). And second, different surviving fractions of the old flora and fauna have withdrawn (died back) either into the southeastern corner of Australia with Tasmania or into subtropical and tropical northern New South Wales and Queensland. Decreasing rainfall may have had more to do with this than temperature. The record of fossil plants does *not* show a whole tropical flora retreating northward in Australia as the climate became cooler in the late Tertiary and Pleistocene. Rather, it shows that the southern Tertiary flora was decimated and dismembered, I think mainly by increasing aridity, and that the dismembered parts tended to be sorted into subtropical and south-temperate groups. The sorting presumably depended partly on the different temperature tolerances or requirements of different groups, but was probably accentuated by aridity. Relative dryness of the intervening area probably forced many plants and animals to make, or seem to make, a more decisive choice between the subtropical and southern cold-temperate areas than they would have had to make if the whole of eastern Australia had been equally wet.

An example of sorting rather than simple retreat is found in *Nothofagus* itself. Three groups of the genus were represented in southern Australia in the Tertiary (see page 144). (No such diversity of groups of *Nothofagus* has ever existed in any really tropical climate, so far as is known.) Since then, one of the groups has disappeared from the whole of Australia (and from New Zealand) and has been "sorted" into the tropics of New Guinea and New Caledonia, while the other two groups have been "sorted" into Tasmania and scattered localities in southeastern Australia. "Sorting" is suggested also by some Carabidae, including *Trichosternus* and the Migadopini (above), which apparently occurred together in the forests of southern Australia in the Tertiary but have died back in opposite directions.

In summary of the Tertiary: within the limits of southern Australia and Tasmania, the main history of the biota during and since the Tertiary has been one of decimation and retreat caused at least partly by decrease of rainfall, and of "sorting" of surviving groups partly according to their temperature preferences. This, of course, has been the history of the plants and animals of the wetter country. Dry-country forms have presumably increased and spread recently. This main pattern of movement in Australia has been complex enough, probably far more complex than we know, but it is only one level of complexity in the history of dispersal of plants and animals in the Australian Region. Within it is a much finer pattern of endless local movements and local speciations. Keast's (1961) study of bird speciation in Australia is one contribution to this subject. And the Tertiary pattern, of decimation and retreat of humid-country organisms and increase and spread of arid-country ones in Australia, is in turn only part of a still wider pattern of evolution and dispersal over the whole world. That some of the retreats in Australia have gone against the direction of the apparent world-wide dispersal pattern adds complexity to the latter but is not evidence against it. Retreats—partial extinctions—are not the same as dispersals and need not go in the same direction, and some countermovement is expected in any case.

The analogy of the tidal river is useful again here, a river of a scale and complexity so great that it is hard to see as a whole. The biotic tide has been receding in the wetter parts of Australia and has left accumulations of relicts of rain-demanding plants and animals especially in the southeastern and northeastern corners of the continent. But the main river of dispersal has been flowing into Australia at the same time, bringing with it a variety of terrestrial vertebrates (Darlington 1957:572–573), *Bembidion* (p. 47), other Carabidae and other insects too numerous to list here, and I suppose also plants, although I am not enough of a botanist to name them.

10. History of New Zealand

The geologic, climatic, and biotic history of New Zealand (Fig. 18) is known in much more detail than the history of any other southern cold-temperate land. Fortunately this history has been summarized by an experienced and restrained New Zealand paleontologist, Charles A. Fleming (1962), to whom I am indebted for additional information and corrections. I have also profited from discussion of New Zealand biogeography with Dr. H. B. Fell.

Geologic-geographic history. An outsider's first reaction to Fleming's survey of New Zealand's geologic history is to be overwhelmed by the complexity of it. New Zealand may in fact have had a more complex history than southern South America or southern Australia and Tasmania, but the histories of these less-studied areas may have been more complex than is yet known. However, in spite of its complexity, the geologic history of New Zealand can be reduced to three principal stages.

During the late Paleozoic and Mesozoic, New Zealand lay along a broad trough (geosyncline) some one or two hundred miles wide and thousands of miles long (Fleming, pp. 58, 68). The trough apparently extended north to New Caledonia and perhaps farther, and there may or may not have been a continuous ridge of land in that direction or a line of close-set islands. The evidence does not show that the trough extended south to the antarctic (Fleming, pp. 59, 61), and an extension of land far in that direction is less likely.

During the Tertiary, the geologic and geographic pattern of New Zealand changed to one of folds, welts, and troughs on a finer scale, the welts tending to be submarine ridges or small areas of land, so that we can think of New Zealand in the Tertiary as an archipelago, as it still is, although details have changed so often and so complexly as to suggest "writhing of part of the mobile Pacific margin" (Fleming, p. 68). Continuous land bridges are not likely to have extended far in any direction during this

time. Moreover, terrestrial mammals, frogs (other than archaic
Leiopelma on New Zealand only), snakes, and other animals that
would almost surely have crossed a continuous bridge if one had
existed are absent on New Caledonia as well as New Zealand
(Darlington 1957:525, 535–536), and their absence seems to guar-
antee that both islands have been isolated from New Guinea and
Australia at least through post-Mesozoic time and probably longer.

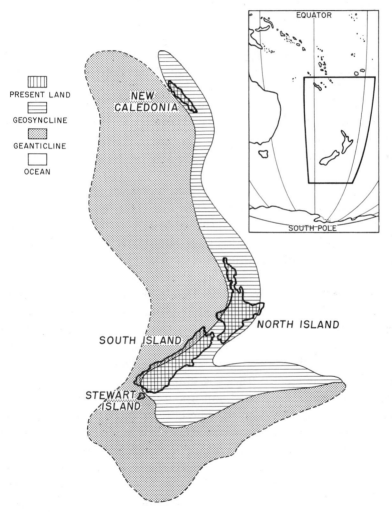

Fɪɢ. 18. New Zealand and New Caledonia: geographic and geologic relations (from Fleming 1962, Fig. 3, redrawn and modified).

Zoogeographic evidence indicates also that New Zealand has been isolated from New Caledonia for a long time. New Caledonia lacks all the old, distinctive, flightless terrestrial vertebrates of New Zealand: the Tuatara, *Leiopelma* frogs, viviparous geckos, and moas and other flightless birds. And the two islands have very different, obviously old, endemic faunas of insects including different Carabidae, for example many flightless Broscini on New Zealand but not on New Caledonia, and flightless Scaritini, including very distinct endemic genera on New Caledonia but not on New Zealand.

New Zealand was apparently not glaciated in the Carboniferous or the Permian. At least no evidence of glaciation then has been found in marine sediments, which indicates that, if ice occurred on land, it did not come down to sea level. Moreover, the two most characteristic fossil plants associated with Permo-Carboniferous glaciation elsewhere in the Southern Hemisphere, *Glossopteris* and *Gangamopteris,* have not been found on New Zealand (Fleming, p. 60).

During the Pleistocene, the mountainous western part of South Island was glaciated at times (Fleming, map on p. 89), but the rest of New Zealand was nearly ice-free. Also during the Pleistocene, eustatic changes of sea level alternately connected and disconnected the different islands of New Zealand.

Climatic history. Because New Zealand, like southern South America and Tasmania, has an oceanic climate closely related to the temperature of the surrounding sea, evidence from shallow-water marine fossils as well as from terrestrial life can be used in deducing past climates on the land, and Fleming has used both sources of evidence to trace the climatic history of New Zealand. The climate has fluctuated, but apparently (until the Pleistocene) only within moderate limits. The climate of New Zealand at the beginning of the Permian was apparently not glacial even when southern Australia was heavily ice-covered (see above). The marine and terrestrial fossil record on New Zealand through the Mesozoic and Tertiary suggests the interplay throughout this time of two climatic elements, one cool-southern, the other tropical or subtropical (Fleming, p. 60). After a warmer period earlier in the Mesozoic, the climate was apparently cooler in the Cretaceous and very early Tertiary, warmer in the mid-Tertiary, then cooler in the Pliocene and cold in the Pleistocene. However, there are indications—such as the absence of tropical reef corals in the Triassic and Upper Jurassic, even though "Tethyan" elements were dominant at the latter time (Fleming, pp. 63, 64)—that the climate of New Zealand was never fully tropical in the Mesozoic and Tertiary, and on the other hand it was apparently never glacial until the Pleistocene.

Biotic history. The earliest fossil plants known from New Zealand are Permian forms mostly like those of Australia (Fleming, p. 59) but not including *Glossopteris* and its immediate allies. Conifers (araucarians and podocarps) appear in the record in the Triassic and Lower to Middle Jurassic (Fleming, p. 63). Some existing groups of animals may have reached New Zealand then, including perhaps the ancestors of the Tuatara and the New Zealand frogs. But these animals may have come later. No fossils have been found to show their time of arrival. Judging from what other reptiles and frogs have done elsewhere in the world, the few that have reached New Zealand may have crossed rather wide water gaps (Darlington 1957:527). It is noteworthy that dinosaurs have not been found fossil on New Zealand (Fleming, pp. 92, 93), although these animals did reach Australia and South America, and although the record of the Mesozoic in New Zealand is fairly good and does include marine reptiles. Angiosperms appeared in New Zealand (as in most other parts of the world) during the Cretaceous, and *Nothofagus* is among the earliest definitely identified, from pollen in the Upper Cretaceous.

The detailed record of the accumulation, evolution, and partial extinction of the fairly rich flora that existed on New Zealand during the Tertiary cannot be traced here. Fleming (pp. 72, 75, 78, 81, 84) gives the main outline of it. He notes (p. 72) that no single direction of immigration can account for the newcomers in the fossil pollen record. Apparently Australian, tropical Malayo-Pacific, and antarctic or subantarctic (Austral) elements all came in. This is said of the Eocene, but it was apparently the case also at other times, judging from Fleming's account (pages listed above) and from his later (1964:373) table of first appearances of various groups of plants in the fossil record on New Zealand. Comparatively little record has been found of the terrestrial fauna. The oldest moa bones are apparently in the late Miocene

(Fleming 1962:81). This is interesting in connection with the evolution of moas but not very important biogeographically, for most
biogeographers and evolutionists now think that the moas and all
the other flightless birds on New Zealand are derived from winged
ancestors that arrived by flight. We (the zoogeographers) find it
difficult to credit a land bridge that would be crossed by birds on
foot but not by other contemporaneous land vertebrates.

Pleistocene climate did not reach its worst on New Zealand
until rather late in that period, although cooling, and geographic
differentiation of the flora on different parts of the island(s), had
begun even in the Pliocene (Fleming, p. 84). The effects of late
Pleistocene climatic changes were profound. Although only the
main mountain range of South Island was extensively glaciated,
much of the rest of New Zealand was reduced to "subalpine"
grassland, with forests apparently (according to the pollen record)
killed back to the northern peninsula and narrow coastal strips
(Fleming, map on p. 89). *Nothofagus* of the *brassii* group, which
had "flourished" through the mid-Pleistocene (Fleming, p. 86),
disappeared from New Zealand in the late Pleistocene, although
other *Nothofagus* and also a rather surprising variety of other
plants survived, as did many insects and other invertebrates,
birds, and the few other, old vertebrates that exist on New Zealand. The re-expansion of forests since the Pleistocene, and other
details of the complex recent history of plants and animals on
New Zealand, are briefly mentioned by Fleming (pp. 90–91, with
references).

Summary of history and significance of New Zealand. New Zealand is
a remote island or group of islands now and it seems to have
been insular throughout the time here considered, with an oceanic
climate which fluctuated but was probably never fully tropical
and never glacial until the Pleistocene.

The evidence of geographic isolation of New Zealand before and
during the Mesozoic is not conclusive, but absence from the fossil
record of significant terrestrial vertebrates, including dinosaurs, is
an indication of isolation, and so is absence of evidence of either
making or breaking of any specific land connection at any specific
time. When a new land connection is made, a massive rush of life
is likely to occur across it, as did occur across the new Central American land bridge at the end of the Pliocene (pp. 77–78), and the

fossil record should show it. The ending of an old land connection ought to show in the record too. But, so far as I can judge, nothing of this sort appears in the record on New Zealand at any time.

Evidence of isolation later, during the Tertiary, does seem conclusive. The nature of the existing fauna, especially the absence of all terrestrial mammals except bats, seems to prove that New Zealand has been isolated from all habitable continents at least since the beginning of the Tertiary, and this is consistent with the geologic evidence given at the beginning of this chapter. Nevertheless the fossil record shows new plants continually appearing during the Tertiary, apparently coming from several different directions. The record therefore leaves little doubt that a variety of terrestrial plants and presumably also invertebrates and some special vertebrates (mostly birds) have reached New Zealand across water from time to time during the Tertiary, just as, somehow, they have reached every other remote island in every ocean. The winds and ocean currents that circle the southern end of the world can account for arrivals on New Zealand from the west and south, from the directions of Australia and Antarctica. And quantities of drift occasionally reach New Zealand from the tropics too (Mason 1961).

I think the evidence is strong that all the plants and animals that have reached New Zealand have done so across water. This seems to me certain in the Tertiary and probable before that. The significance of this fact, if it is accepted as a fact, goes far beyond the limits of New Zealand itself. Most of the peculiar southern cold-temperate groups of plants and invertebrates that are common to southern South America and southern Australia-Tasmania occur on New Zealand too. For example, of 62 "south-temperate amphitranspacific" groups of plants listed and discussed by van Steenis (1962:256–265, 267), all except 2 or 3 are represented on New Zealand. If these plants and all the other plants and animals that have reached New Zealand crossed water gaps, no land connections are needed anywhere across the southern end of the world to explain the distribution of far-southern terrestrial life.

For further discussion of the origins of the fresh-water fishes of New Zealand, see McDowall (1964), and for zoogeographic analysis of the whole terrestrial vertebrate fauna, see Caughley (1964).

11. History of South Africa and India

South Africa. The latitude of the southern tip of Africa is only about 35° S. The climate there is *warm*-temperate. And the flora and fauna include very few southern cold-temperate elements. Nevertheless South Africa should be considered briefly, if only to justify dismissing it.

The geologic and climatic history of South Africa need not be given in much detail here. The land is old. The whole of southern Africa north into what is now the edge of the tropics (about 22° S) was apparently heavily glaciated mainly in the late Carboniferous (King 1962:40–42, Fig. 16). Southwestern Madagascar was apparently glaciated too, and the *Glossopteris* flora was represented there as well as in South Africa during or after glaciation. Direction of movement of the ice on South Africa was mainly toward the south and southwest (if the continent was in its present position) but some details are doubtful (see p. 183). No glaciation occurred on South Africa in the Pleistocene, not even on the highest land (Brink 1960:568).

The existing flora of South Africa is very rich and is in some ways similar to the flora of Australia. Hooker pointed this out long ago (see Turrill 1953:181–182). The similarity is mainly a matter of orders and families that are dominant in both places. Relatively few genera occur in both. And the groups that are shared are not special southern cold-temperate ones. Notably, *Nothofagus* does not occur in South Africa and never has occurred there, so far as is known (Levyns 1962).

Some "possible antarctic elements" in the South African flora are discussed by Levyns. They are of great interest and are worth remembering and rediscussing now and then, but they need not have had anything to do with the antarctic. I cannot review every case but shall comment briefly on the southern conifers as examples, because they are favorites with phytogeographers.

South Africa has only two existing indigenous genera of conifers. One, *Podocarpus,* is represented by a concentration of 5 species in southern Africa (Levyns), but this genus occurs in tropical

Africa too and is widely distributed on all tropical continents as well as in the south temperate zone (Florin 1963:199, Fig. 25; my Fig. 33). It may as well have dispersed through the tropics as by antarctic routes. The other existing South African conifer genus is the endemic *Widdringtonia,* which occurs in South and East Africa north well into the tropics, to about 15° S (Florin, pp. 236–237, Fig. 48). The present relationships as well as the fossil record of this genus are apparently doubtful. Its supposed Australian relative *Callitris* (Florin doubts a close relationship) does not occur with *Nothofagus* but lives in different, warmer habitats: in the opener, warmer eastern part of Tasmania and north into the tropics on the Australian mainland (*Australian Encyclopaedia*). A third conifer genus, *Araucaria,* is fossil in South Africa in the Lower Cretaceous, but this genus is subtropical rather than antarctic in present distribution and is apparently world-wide in the fossil record (Florin, pp. 176–180, Fig. 14; my Fig. 32). None of these conifer genera is "antarctic" in the way that *Nothofagus* is.

The vertebrate fauna of South Africa includes one representative of a truly southern cold-temperate group, a galaxiid fish, which lives in fresh water but probably came through the sea (see p. 38). Some galaxiids do enter the sea. And it is difficult to design a land bridge that would carry a fresh-water fish to Africa without carrying terrestrial vertebrates too. One genus of frogs (*Heleophryne*) in mountain brooks in South Africa is usually considered a leptodactylid, and leptodactylids now occur otherwise only in Australia (with New Guinea) and South America (with Central America and the West Indies), but the family is more tropical than southern cold-temperate in distribution and is supposed to be fossil in India in the Eocene (Darlington 1957: 166).

A few South African invertebrates belong to "antarctic" groups (Brink 1960:571). They live on mountains in characteristic habitats, "in ravines, wet heaths or forests, in mountain streams or in wet moss covering rocks," but not in other habitats and notably not in the very ancient Namib Desert. Stuckenberg (1962) describes and maps the distributions of some of the South African insects and Onychophora in question. His paper, and the one by Levyns referred to above, give useful details of distributions in South Africa, but these papers would be still more useful if the climatic as well as the geographic ranges of pertinent groups outside Africa were specified. For example, it might be useful to biogeographers to know that stag beetles of the subfamily Lampriminae (Stuckenberg, p. 191) are not peculiarly southern cold-temperate insects but are represented also in warm-temperate parts of South America and Australia and in tropical New Guinea, as well as (as Stuckenberg says) by a fossil in Tertiary amber in northern Europe.

A few of the special plants and animals that now live in South Africa may conceivably be relicts of ancient antarctic stocks, but other explanations of their distributions are possible. Florin (p. 237) suggests that all the conifers that have reached Africa may have come from the northeast, and Brink thinks that the "antarctic" insects and other invertebrates in special habitats on mountains on the southern tip of Africa may be relicts of invasions from the north or may have been dispersed by winds and ocean currents rather than over land connections in the south.

Climate has important effects on distribution of plants and animals in South Africa. Although the southern tip of Africa is not cold, it is climatically differentiated and also is isolated from the main part of the continent by a barrier of aridity that evidently retards southward dispersal of water-demanding organisms. This situation favors persistence of relicts in wet habitats (where the relict invertebrates do in fact occur, according to Brink) on the southern tip of Africa, regardless of the direction from which their ancestors arrived.

In summary: the supposed "antarctic" plants and animals on the southern tip of Africa are relatively few. Some (most?) of them do not belong to true antarctic or southern cold-temperate groups. Their present geographic relationships are complex. Their geographic origins are doubtful. And their present distributions are correlated with present climate regardless of what their geographic history has been. They pose an interesting biogeographic problem, but I think it is a secondary one. Solution of it may follow solution of the primary problem of the history of life at the southern end of the world, but I think the situation in South Africa contributes little to solution of the primary problem of far-southern biogeography. This is my justification for not treating South Africa in more detail here.

India. Although India now lies north of the equator, the distribution of glaciation and of floras late in the Paleozoic suggests an association with the southern continents in the past. This is at least compatible with the known paleomagnetic record, although the latter is still very unsatisfactory so far as India is concerned (see p. 203).

Peninsular India is a very old piece of land, much of it based on Precambrian rocks (Kummel 1961:83, Fig. 4–8), and this part of India has been remarkably stable and mostly above sea level since the later Paleozoic, with only minor, brief invasions of the sea in the early Permian and in the Cretaceous (Kummel, p. 329). In the past, this part of India has been separated from Asia by the Tethyan geosyncline, a geologically active zone of fluctuating seas and, later, mountain formation. Northern India, including what is now West Pakistan, was apparently involved in the history of the Tethys from the Permian through the Mesozoic (Kummel, pp. 155, 264–268). I do not know whether this involvement can be reconciled with the movement of India postulated by Wegenerians.

India was extensively glaciated late in the Paleozoic, presumably at the same time that glaciation occurred in the Southern Hemisphere. Evidence that ice sheets existed on India is unmistakable, but authorities disagree about the direction of movement of the ice. Kummel (p. 334, Fig. 11–6) and Schwarzbach (1963: 140–141, Fig. 87) accept one pattern of movement, and King (1961:310, Fig. 1), a fundamentally different one.

A *Glossopteris* flora is fossil on India in late Paleozoic glacial and postglacial deposits. Related floras are fossil in the Southern Hemisphere wherever glaciation occurred. The relationships between the Indian and far-southern floras were apparently very close. For example, Plumstead (1962:68) considers 28 *species* common to the Permo-Carboniferous floras of India and Antarctica, and these apparently identical species constitute more than 70 percent of the known, identifiable Permo-Carboniferous plants of Antarctica! The specific identities are necessarily based principally on superficial leaf form and may be doubtful, but the remarkable similarity of the Indian and Antarctic floras as wholes can hardly be doubted. Fossil tree trunks with pronounced annual rings have been found with these floras in India (Seward 1931:244–245, Fig. 73) as well as in the Southern Hemisphere, suggesting that the climate of all the glaciated areas was not only cold but also seasonal, probably with well-marked alternation of summers and winters, which would be expected only at high latitudes.

The later history of India cannot be considered in detail here. Significant facts are few and in part doubtful and confusingly inconsistent. For example Florin's (1963:265, Fig. 64; my Fig. 34) summary map of the distribution of fossil conifers seems to show that, floristically, India was associated with the southern continents from the late Carboniferous to the Eocene, but Florin's own detailed maps raise doubts about this. The detailed maps show araucarians widely distributed in the Northern Hemisphere as well as on the southern continents and India (Florin, p. 178, Fig. 14), and all the fossil Podocarpaceae in India are questioned (Florin, pp. 182–200, Figs. 17–25). Land-living reptiles have been found fossil in two "Gondwana" horizons in peninsular India. "The reptiles of the lower horizon are forms characteristic of South Africa and northwestern China. The fauna of the upper horizon is like that of the Upper Triassic of the northern hemisphere" (Kummel 1961:340–341). Nevertheless paleomagnetism seems to place India very far south of its present position, about halfway between the equator and the South Pole, as late as the Jurassic. Can these conflicting indications be reconciled?

12. History of Antarctica

Recent publications on Antarctica are so numerous and diverse as to defy a full review and I have not attempted one, although I shall refer to the latest publications that I have been able to find on critical details. (See Robin 1964 for references to reports on current antarctic research.) The two-part report on "Science in Antarctica" by the Committee on Polar Research of the National Academy of Sciences–National Research Council (1961) is a useful progress report on many aspects of antarctic research. A less technical account (unfortunately without references) of what is now known of the antarctic region, its history, and its life will be found in a set of articles in the *Scientific American* Antarctic Issue (1962). That there are a few errors in so broad and useful a work is excusable. (I hope I may make as few, and be forgiven them!) An error that ought to be corrected is the statement on page 213 that the southern limit of trees in the Southern Hemisphere is the Beagle Channel. This statement is not correct and it invites a wrong inference as to how different the climate would have to be to permit trees to grow on Antarctica. Actually, heavy forest of large *Nothofagus* trees now exists on northern Navarino Island south of the Beagle Channel, and stunted forest occurs still farther south (see p. 30).

Geologic-geographic history. Geologically, Antarctica (Fig. 19) consists of two parts (Hamilton 1963). The eastern part, toward Africa, is a very old, relatively stable continental shield. The western part is a relatively unstable area in which mountains have been formed at various times, in part (on the Antarctic Peninsula) as recently as the Cretaceous or early Tertiary (Hamilton, p. 7). This part of Antarctica may now be an archipelago rather than continuous land. It includes the Antarctic (or Palmer) Peninsula, which may really be an island (Euller 1960:158, map; Behrendt and Parks 1962).

A continental ice cap now covers virtually the whole of Antarctica, joining the separate parts. The weight of the ice has caused

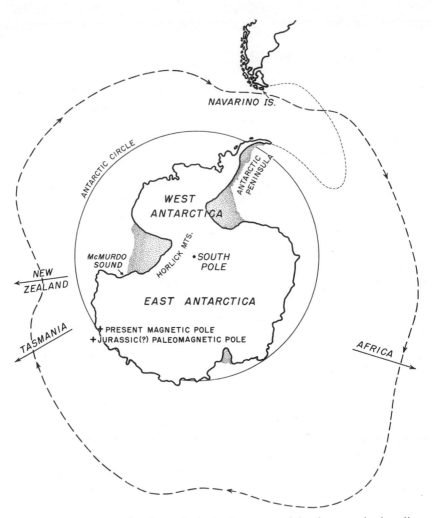

Fig. 19. Antarctica, showing principal places named in the text; broken line with arrows is main course of the Circumpolar Current, simplified (for further details see Kort 1962: 116, map).

the whole continent to sink below the level it would otherwise have. Rock is uncovered only on mountains that rise above the ice level and in widely separated, small, ice-free areas along the coast. The existing ice makes it very difficult to see what the surface of the land is like or what its history was before the ice formed. However, evidence of late Paleozoic glaciation has been found recently on an exposed escarpment in the Horlick Mountains (Doumani and Long 1962:172), and this suggests an early continental glaciation contemporaneous with the Permo-Carboniferous glaciations of other southern continents. This suggestion is supported by the presence of fossil *Glossopteris* above the old glacial deposits (Doumani and Long, p. 175), and *Glossopteris* has long been known from other localities on Antarctica (Barghoorn 1961; Plumstead 1962:11, map).

The western part of Antarctica, which points toward South America, is considered an extension of the Andean mountain system. The latter is geologically active and has risen mainly during the Tertiary, and a complete but narrow and sinuous land connection, an isthmian link, may have been formed at times between Antarctica and South America. There is, however, no actual proof of this. Proof is not likely to come from the distribution of terrestrial plants and animals because of the extreme poverty of the Antarctic land biota. Perhaps the record of distribution of shallow-water marine animals will prove or disprove the existence of a connection (see pp. 120–121).

A curious detail of geologic history is that the south magnetic pole at some time probably in the Jurassic (or possibly the Cretaceous) was almost where it is now in relation to Antarctica (Turnbull 1959; Woollard 1962:164–166, map on p. 152). Nevertheless paleomagneticists think that the south pole of rotation then was perhaps 25° or 30° from where it is now. And this conclusion is legitimate, if paleomagnetic theory is correct. The south magnetic pole is supposed to wander randomly around the true south. At any one moment of geologic time the magnetic pole is likely to be some distance from the true south, as it is now. In fact the south magnetic pole has moved rapidly recently, about 800 km in the last 50 years (Woollard, pp. 164–166). But the position assigned the south magnetic pole in the Jurassic is an average of many determinations from rocks that represent a considerable interval of time, and if the theory is correct, the average of all determinations should approximate the true south. This principle is so important that it is worth clarifying by an analogy. The average of many readings of temperature taken at Boston at various times on March 1 should be about 33°F. A single reading made on May 1 may happen to be 33° too. But this does not mean that May is the same as March in temperature. Similarly, if the average Jurassic postion is almost the same as the present single position of the magnetic pole, this does not mean that the south pole of rotation was the same.

It should be noted that, even if the Jurassic pole was where pa-

leomagnetism indicates, one edge of Antarctica was still within the
Antarctic Circle and no part of the continent was in the tropics.

Climatic and biotic history: ancient glaciation and fossil floras. Something of the climatic history of Antarctica can be inferred from
fossil floras in conjunction with some other evidence. So little is
known that there is not much point in separating climatic and
biotic histories. The two are interdependent. All the information
together is very limited, and the limits should be recognized.

The climate of Antarctica in the late Paleozoic was apparently
at times glacial (see above) and later, in the Permian, apparently
humid cool-temperate. This is indicated by (for example) the finding of coal beds as much as 13 feet thick on the Horlick Mountains
(Doumani and Long, p. 175) and by the occurrence of coal and
probable *Glossopteris* leaves and conifer needles scarcely more than
3° from the present south pole (Barghoorn 1961).

Fossil plants have been found on Antarctica in the Devonian,
Permo-Carboniferous, Triassic, Jurassic, (late?) Cretaceous, and
part of the Tertiary (Plumstead 1962, with map on p. 11). The
Permo-Carboniferous flora is the largest and most widely distributed one that has been found on Antarctica. It consists mainly of
Glossopteris and allied genera, "and very few other plants occur"
(Plumstead). Fossil wood found with this flora is from trees of considerable girth and shows annual rings indicating a seasonal climate. This flora on Antarctica is almost always associated with
coal seams. There is little direct evidence that it was associated
with glaciation on Antarctica, but *Glossopteris* is often found with
actual glacial tillites on other continents (Plumstead, pp. 30, 67),
and other evidence suggests that floras like this could grow and
form coal in polar or subpolar regions (see p. 133). This species-
poor flora (species-poor as compared with floras in some other
parts of the world but still apparently the richest yet found on
the main part of Antarctica) indicates nothing like a tropical climate and probably does not require a change of position of Antarctica.

Very few Triassic and Jurassic plants have yet been found on
the main part of Antarctica (Plumstead), but a moderately diverse
Jurassic flora has been found on the Antarctic Peninsula (Barghoorn, p. 6), although coal has not been found. These plants are
sometimes said to indicate that Antarctica had a warm or even

tropical climate, but this is doubtful. The climatic tolerances of the Jurassic plants are unknown. Moreover, they existed on a peninsula (or archipelago?) perhaps 1000 km (600 miles) out from the Antarctic mainland, in what may have been an oceanic climate perhaps much milder than the mainland climate.

Some botanists think that angiosperms originated in the Jurassic and radiated on and from Antarctica, but no trace of them has been found in the floras referred to above. Plumstead (p. 85) notes that the few Jurassic plants that she examined from the mainland of Antarctica are all gymnosperms; and the larger, probably Upper Jurassic, flora of the Antarctic Peninsula summarized by Barghoorn includes no angiosperms. This latter flora, incidentally, is said by Barghoorn (p. 6) to be most closely related to the Lower Oolite flora of Yorkshire, England, and "quite comparable" with fossil floras from the Upper Gondwana series of India.

Good samples of Upper Cretaceous and Tertiary land floras have been obtained only near the tip of the Antarctic Peninsula. The deposits are actually marine, but the plant fragments and pollen in them probably came from nearby land (Cranwell, manuscript). They include araucarian pines, all three groups of *Nothofagus,* and other plants indicating (I think) a forest growing under moderate-temperate, but not tropical, conditions. This forest apparently existed on the Antarctic Peninsula from the Upper Cretaceous through much of the Tertiary—the latest records of it are apparently Miocene. If the Antarctic Peninsula was an island during this time (see above), the forest may have occurred only on an archipelago with climate tempered by the sea and not on the Antarctic Continent itself, except perhaps along its margins. I do not mean that this was the case but only that the record from the peninsula does not prove that forests were widely distributed on Antarctica or that the climate of the interior was suitable for them.

Small amounts of pollen "at low frequencies—only a few grains per slide" of *Nothofagus,* conifers, and a few other plants have been found in morainic material probably of early Tertiary age dredged from the sea bottom in the McMurdo Sound District off Antarctica at about 166° E (Cranwell, Harrington, and Speden 1960). Only small, wind-pollinated types are present, and pollen of this sort is known to be blown very great distances (see p. 142) and may even be reaching Antarctica now (Holdgate 1964). Nevertheless the McMurdo pollen does suggest the presence of forest including *Noth-*

ofagus on the mainland of Antarctica. This is not inherently unlikely, but in view of some past exaggerations it might be wise to
remember that the evidence in this case is not conclusive.

History of land animals on Antarctica. Almost no fossil record of land
animals has been found on Antarctica. Tracks apparently made by
vertebrates have been reported in Permian strata on the Ohio
Range of the Horlick Mountains, but the distinctive, terrestrial
and aquatic reptiles associated with "Gondwana" deposits in
Africa and South America have not been found on Antarctica
(Doumani and Long, p. 176). This is negative evidence which may
or may not be significant. Doumani and Long suggest that mesosaurs
may turn up in the Permian of Antarctica. These animals were
swimming reptiles, known fossil in early Permian fresh-water or
estuarine deposits in South America and South Africa. Like some
existing turtles and crocodiles, they may have been able to disperse
by sea. Their discovery in Antarctica (they have *not* been discovered) would be an interesting event but would hardly imply land
connections. The discovery of terrestrial vertebrates fossil on Antarctica, especially of dinosaurs in the Mesozoic (they have *not* been
discovered), would be much more exciting, although even their
significance might be doubtful. If dinosaurs ever did reach Antarctica, the remains of some of them might be expected to have
washed into the sea and to be found fossil in shallow-water deposits there. I do not know whether enough such deposits have been
explored to make failure to find dinosaurs significant. So far as I
know, the only presumably nonmarine invertebrates found fossil
on Antarctica are fresh-water Crustacea of "Gondwana" types
from the Middle-Upper Permian of the Ohio Range of the Horlick
Mountains (Doumani and Tasch 1963), unidentified fragments of
insect wings in the Permo-Carboniferous of the mainland of Antarctica (Plumstead 1962:66–67), and two beetles from the Jurassic
of the Antarctic Peninsula (Zeuner 1959).

In spite of the dearth of fossils, something more can be said of
the occurrence of invertebrates on Antarctica in the past. Forest
has existed on Antarctica. I think it is true of all existing forests
that they are always inhabited by insects, even on the most remote
islands and in the most adverse climates where forests occur. Insects and other appropriate invertebrates probably lived in the
forest on Antarctica, and moorland as well as forest probably ex-

isted there, with moorland invertebrates. And some of the insects and other invertebrates on Antarctica in the Tertiary probably belonged to groups that now survive in forest and on moorland in the small, widely separated land areas in the southern cold-temperate zone. Some vertebrates too that are able to disperse across gaps of salt water, for example land birds and probably galaxiid fishes, presumably occurred in the forested parts of Antarctica in the Tertiary.

Antarctica surely has had a significant flora and probably also a considerable terrestrial fauna at times in the past. What part these plants and animals may have played in evolution and dispersal of existing floras and faunas in the southern cold-temperate zone is considered in Chapter 17, under the heading "Role of Antarctica." The history and significance of antarctic marine life is briefly discussed in Chapter 19, under "Shallow-water marine faunas."

Present climate and biota. The latest fossil terrestrial floras even on the Antarctic Peninsula are apparently Miocene. Antarctica probably cooled in the Pliocene and became ice-capped in the Pleistocene, if not before. The cooling was approximately contemporaneous with cooling of the northern end of the world and was evidently the result of a great revolution of climate, not of great changes in position of the continents. The paleomagnetic record of the northern continents (the record of the southern ones is inadequate) shows them virtually motionless while the climate changed.

Antarctica may be said still to be in the Pleistocene, with a climate extremely inhospitable to terrestrial life. Mean temperature everywhere is well below freezing and is below $-55°$ C $(-67°$ F) in the coldest part of the high interior. The continent is also rather dry. Precipitation, in water equivalent of snow, is nowhere over 55 cm (about 22 in.) and in much of the interior is under 5 cm (about 2 in.) a year. Winds circle the continent from west to east, but storm winds often turn inland. For more information about present climate see Rubin (1962, especially maps on pp. 88–89).

Under these conditions, very few terrestrial organisms have succeeded in reaching or surviving on Antarctica. No trees occur there and only three species of small flowering plants (one herb and two grasses), and they are confined to the peninsula. The only land plants that now tolerate the climate of the main part of Antarctica are algae, lichens, and mosses. Sea birds and marine mammals live around the edges of Antarctica but no terrestrial vertebrates occur there. Of true insects (other than ectoparasites of the birds and marine mammals) only two species of flies reach Antarctica, and they actually reach only the peninsula or islands off the peninsula, although Collembola and free-living mites occur on parts of the Antarctic Continent proper. For description of existing land life of Antarctica see Llano (1962), and for additional information on some of the terrestrial arthropods see Gressitt (1964b) and Gressitt, Leech, and Wise (1963), who state that the southernmost known locality for arthropods or for any permanent free-living animals on Antarctica now is the vicinity of Hood Glacier, 83° 55′ S. The plants and animals that do exist on the Antarctic Continent, limited though they are, form a series of zones or "life-spheres" in which diversity of life gradually increases from the center of the continent outward (Gollerbakh and Syroechkovskyi 1960). (In comparing different accounts of the occurrence of animal life on Antarctica, readers should understand that Collembola are still classified as insects by some persons while I and some other entomologists now prefer to treat them as a separate group of arthropods.)

It should be emphasized that a moderate improvement of climate and melting of the edges of the ice cap would probably allow forests and other vegetation and associated animals to occupy the Antarctic Peninsula and perhaps parts of the continent proper. Forests now occur in northern Europe and northwestern North America at latitudes equivalent to the latitude of the Antarctic Peninsula, in spite of the fact that the northern climates are much colder now than they have usually been in the past. The position of Antarctica does not forbid existence of forests in the past, and evidence that forests existed is not evidence of change of position of the continent.

The effect on climate of a possible isthmian connection between Antarctica and South America has probably been considered before now, but I have not come across a discussion of it. Presumably such a connection would change the circulation of the Antarctic Ocean. Instead of flowing around the continent as it does now (Kort 1962:maps on pp. 114, 116; my Fig. 19), the main ocean current in the far south would probably be turned up the west coast of South America, and warm water might flow down the east coast in greater quantities and farther than now. The expected result would be to increase the difference that exists even now between the colder climate of the west and the warmer one of the east coast of southern South America and to extend the difference southward, on opposite sides of the hypothetical connection, perhaps even to adjacent parts of the coast of Antarctica. If such a situation existed at times in the past, evidence of it should be found in the distribution of fossil terrestrial and shallow-water organisms.

Summary of history of climate and vegetation of Antarctica. The actual evidence of past climate and vegetation of Antarctica can, I think, be summarized in three parts. (1) In the Permo-Carboniferous, although actual glacial deposits have been found only very locally, a species-poor, coal-forming flora that is associated with glaciation elsewhere was widely distributed on Antarctica and suggests a cold-temperate, seasonal climate. (2) At times in the Mesozoic, notably in the Jurassic, some plants grew on at least the edges of the mainland and a moderately diverse flora grew on what is now the Antarctic Peninsula, but its climatic requirements are not known, and the nearness of the south magnetic pole in the Jurassic suggests a less than tropical climate. (3) And in the late Cretaceous and Tertiary, a temperate flora, including *Nothofagus,* grew on the Antarctic Peninsula and perhaps also on parts of the Antarctic Continent proper.

This evidence warrants the conclusion that Antarctica was warmer than now in the Permian (after glaciation), Mesozoic, and Tertiary. Temperate forest may have occurred on parts of Antarctica much of the time, although both climate and vegetation probably varied greatly from place to place on the continent, as they do on all other continents now. But there is nothing whatever in the record I have summarized to justify the widely held belief that Antarctica has been tropical or subtropical. The evidence is against it in detail, and the general probability is against it too. Tropicalness requires a uniformity of temperature throughout the year that is impossible at high latitudes, and the Antarctic Continent seems to have been somewhere near the South Pole for a very long time.

It is probably true that no organism known to have occurred on Antarctica is a sure indicator of land connections. In view of the numbers of plants and animals that have reached New Zealand apparently across water barriers during the Tertiary and that have reached remote oceanic islands elsewhere, it seems a bit dogmatic to demand land connections to Antarctica for biogeo-graphic reasons, although some persons do so. I perfer to leave the question of antarctic land connections unanswered, at least for the time being.

The history and role of Antarctica are further considered in parts of Chapters 17–21.

13. Summary: similarities of far-southern land areas

Excluding South Africa (because it is not cold-temperate) and Antarctica (because it is ice-covered), the southern lands discussed in Chapters 8–12 are remarkably similar in several ways. These lands are southern South America (with Tierra del Fuego), the southeastern corner of Australia (with Tasmania), and New Zealand.

First, these three land areas have in common a general similarity of position and dimensions. They all extend into or lie in cool latitudes south of the tropics. And they are all either relatively narrow peninsulas or islands, much exposed to the climatic influences of the sea.

Second, these three land areas have in common a considerable degree of isolation, both now and in the past. South America and Australia have, as wholes, been cut off from other habitable continents by water gaps at least since the beginning of the Tertiary and perhaps longer, until South America became linked to North America late in the Pliocene. And the wet southern corners of South America and of southeastern Australia with Tasmania are and probably long have been further isolated by barriers of relatively arid country. New Zealand is and probably long has been even more isolated, by wide ocean barriers. However, for cold-adapted and water-requiring plants and animals coming from the north, southern South America and southern Australia may have been almost as difficult to reach as New Zealand.

Third, these three small southern land areas seem to have had similar climatic histories, at least for a very long time. Whether or not they were all glaciated late in the Paleozoic (Tasmania was, New Zealand apparently was not, and the southern tip of South America may not have been), since then all of them seem to have had generally similar climates. All seem to have been warmer than now but not tropical in the Mesozoic and much of the Tertiary, although the record is of course very incomplete. And, finally, all were glaciated to some extent in the Pleistocene and evidently had Pleistocene climates more severe than now.

Fourth, the three small land areas in question have similar climates now. They are all cool-temperate, with areas of heavy rainfall, and with oceanic climates. This last characteristic is especially important. If the shapes of southern South America, Tasmania, and New Zealand were traced on one of the large land masses in the Northern Hemisphere, at north latitudes equivalent to present south latitudes but distant from the ocean, the climates of the traced shapes would be very different from existing southern climates, with generally warmer summers and colder winters, much wider extremes of heat and cold, and (depending on exact position) perhaps less rain or rain only at certain seasons. The nearness of great expanses of ocean in the far south moderates seasonal cycles of temperature and makes a relatively uniform climate throughout the year, with heavy rainfall especially along the western edge of each piece of land. The climates of the small land areas in the southern cold-temperate zone are actually more like climates at certain altitudes in the tropics than like north-temperate climates (see pp. 89, 131), and this similarity has evidently had important effects on plant and animal distribution.

These similarities of position, past and present isolation, climatic history, and present climate of the three principal land areas in the southern cold-temperate zone are surely partly responsible for the remarkable similarities of their existing floras and faunas, regardless of the geographic origins of the organisms concerned.

The Antarctic Continent proper is polar in position and relatively large, but the Antarctic Peninsula lies not far south of the southern tip of South America, and it is a narrow piece of land much exposed to the sea's influence. The geographic isolation of Antarctica—the distance from the nearest continent—is no greater than the isolation of New Zealand. And, until recently, the climatic history of Antarctica seems to have paralleled the climatic histories of southern cold-temperate lands: the climate of Antarctica was evidently glacial in the late Paleozoic, then warm enough to allow growth of forests in the Permian and Mesozoic, and still warm enough in the early Tertiary to allow forests on the Peninsula and probably also on at least the edges of the continent itself. However, finally, while southern South America, Tasmania, and New Zealand underwent only local glaciation in the Pleistocene, with much land remaining ice-free, Antarctica developed a complete ice cap and still has it. The complete, persistent ice cover may be due more to the extent and height of the land than to the position of the continent or to initial coldness. In any case, the present difference between the climate of Antarctica and the climates of southernmost South America, Tasmania, and New Zealand is unusual. Through most of its pre-Pleistocene history Antarctica, at least the edges of the continent and outlying peninsulas and islands, may have had a climate comparable to the climates of southern South America, Tasmania, and New Zealand, and may have shared large parts of the floras and faunas of these places.

PART IV Inferences about the past

14. Role of climate in the past

Relative importance of climate. Generally speaking, the distribution of any terrestrial organism is determined partly by distribution of land and partly by climate. Land plants and animals can occur only on land that they can reach, and only where climate is suitable. The distribution of *Nothofagus*, for example, reflects both the geography of land in the Southern Hemisphere and the distribution of suitable climate there, and this is true whatever the geographic history of the plants may have been. Other factors are presumably important too, probably especially competition, but they will be ignored for the moment. Given the fact that the distributions of both land and climate are important, can either be said to have been more important in determining distributions of plants and animals in the southern cold-temperate zone? I think that at least half an answer can be given to this question.

The answer depends on the fact that many southern cold-temperate plants and animals are involved in amphitropical distributions, in which related forms occur in the north temperate and south temperate zones of the world but are few or absent in the intervening tropics. The closeness of relationship of southern to northern forms varies in different cases. Among the Carabidae discussed in Chapter 4, for example, southern and northern *Bembidion* belong to the same genus, and some southern species belong to northern subgenera even in Jeannel's finely split classification. Among Trechini and Broscini, southern and northern genera are all different but belong to the same tribes. And although the tribe Migadopini is itself confined to the Southern Hemisphere, it is related to a northern tribe with which it forms an amphitropical pattern at a higher-than-tribal level. A similar variation in closeness of relationship occurs among amphitropical plants (Constance 1963). Some actual species of plants occur in both southern and northern temperate zones. Many genera are represented by related species in the south and in the north. Some other plants, including *Nothofagus*, are generically differentiated in the south but are still closely related to northern genera. And others, including I think in general the conifers, show still wider divergences between southern and northern genera. (However, the distribution of conifers is not strictly amphitropical; see pp. 186–187.)

All together, a large proportion of the plants and animals of the southern cold-temperate zone are involved in amphitropical distributions at one taxonomic level or another, and they include many of the most characteristic southern groups that are supposed by some persons to have dispersed by means of continental drift or antarctic land bridges. But the amphitropical pattern cannot be primarily a product of ancient geography. No biogeographers would seriously suggest (or would they?) that existing north and south temperate areas once formed a single land mass entirely separated from existing tropical areas, and that the amphitropical groups of plants and animals are still distributed according to the ancient division of land. Climate apparently must be primarily concerned in formation of the amphitropical patterns. Moreover, if the differences in level of relationship of southern to northern forms reflect differences in time of dispersal, climate has been concerned in zonation of southern as well as northern plants and animals for a very long time. It does not necessarily follow that climate has been more important than ancient geography, although this conclusion might almost be justified by the prevalence of amphitropical patterns. But it can be said—this is the half answer to the question asked above—that climate, especially zonation of climate, has apparently been profoundly important for a very long time in determining patterns of distribution in the far south. Biogeographers who do not understand the importance of climate cannot understand the distributions of southern plants and animals and are likely to form unnecessary hypotheses of continental drift or ancient land bridges.

Patterns of zonation of climate. Amphitropical distributions emphasize one sort of zonation of climate: differentiation of temperate zones in the Northern and Southern Hemispheres, separated by the tropics. This pattern may be designated *temperate/tropical/temperate.* Zonation of this sort is obvious and has surely determined the present distributions of many southern plants and animals, and has also impeded dispersal of cold-adapted organisms between the Northern and Southern Hemispheres. However, another sort of zonation of climate exists and is important too in a less obvious way. It is differentiation of a zone of relatively unstable, violently fluctuating, "temperate" (really intemperate) climate north of the tropics, contrasted with the more stable climate of the rest of the world. Zonation of this sort may be designated *north-temperate/tropical + south-temperate.*

This second pattern can be looked at either from the north or from the south. From the north, the significant fact seems to be that prevailing northern continental climates are different from climates anywhere else in the world. From the south, the significant fact is that the oceanic climates of the small pieces of land in the southern cold-temperate zone are in some ways more like climates in the tropics than like north-temperate climates (Troll 1960; see also p. 89).

Biogeographers living in the north temperate zone do not always realize that this second pattern of climatic zonation is reflected in the distributions of many plants and animals. Among plants, for example, tree ferns are very widely distributed in the south temperate zone as well as in the tropics. They even grow near glaciers in New Zealand (Seward 1931:frontispiece). But they are absent in the north temperate zone, or nearly so. Some conifers too show a pattern of *north-temperate/tropical + south-temperate* zonation (see maps in Florin 1963, especially the distribution of *Podocarpus,* p. 199; my Fig. 33). This pattern may formerly have been characteristic of the conifers as a whole. The so-called southern conifers do in fact tend to be relict in the tropics as well as in the south temperate zone (see again Florin's maps). Among animals, the Onychophora (*Peripatus* and so forth) follow this pattern. Their distribution is almost world-wide in suitable tropical and south-temperate continental areas. They occur far southward in southern South America and on Tasmania and New Zealand as well as in the tropics (Brink 1957). But they do not extend into the north temperate zone except very slightly in southeastern Asia.

One probable consequence of the partial similarity of southern cold-temperate climates to climates in the tropics is that southern cold-temperate plants and animals can disperse through tropical or subtropical areas more easily than north-temperate plants and animals can. Another consequence is that biogeographers are likely to mistake the climatic significance of southern plants and animals. The mistake is to suppose that groups now confined to the tropics have always been tropical and that their occurrence as fossils in the far south is proof of a formerly tropical climate there. The erroneous idea that Antarctica has been tropical is based on this kind of mistake. Actually the groups in question may have had distributions that included both tropical and cool southern oceanic climates, as the distributions of tree ferns, some conifers, and Onychophora do now.

In spite of the obvious general importance of zonation of climate in the distribution of southern plants and animals, in particular cases the effect of climate may be difficult to distinguish from other factors, especially from competition. For example, climate has certainly been important in general in the geographic history of *Nothofagus,* but climate need not have been directly responsible for all recent changes of distribution of groups of the genus. Specifically, the *brassii* group of *Nothofagus,* which was once widely distributed in the southern cold-temperate zone but which has now "retreated" into New Guinea and New Caledonia, need not have been killed back by Pleistocene cold. Perhaps, instead, the *brassii* group could not stand competition with other *Nothofagus.* The advantage of the latter may have been greatest during the movements that accompanied Pleistocene climatic fluctuations, which necessitated continual invasion and re-invasion of new ground. If so, the *brassii* group disappeared in the far south because it was replaced by other groups of *Nothofagus,* not because it was directly eliminated by cold.

Climatic significance of southern coal floras. The vegetation of the world was strongly zoned in the late Paleozoic, and the zonation was evidently correlated with climate. A special flora, dominated by *Glossopteris* and allied and associated plants, occurred around the world in the far south (and in India), always in areas that were or had been glaciated. Further details are given in Chapter 19 and Fig. 35. Now, I wish only to indicate the apparent climatic limits and significance of *Glossopteris* and of the southern coal-forming flora of which it was a part.

The *Glossopteris* flora lived in a glacial or postglacial climate. To the north, it was presumably limited by warmer climate. Physical barriers adequate to prevent northward spread are unlikely and unnecessary. All the continents seem to have been reasonably accessible then (see p. 83); *Glossopteris* was apparently wind-dispersed (p. 193) and probably able to cross most non-climatic barriers; and the distribution of glaciation as well as of floras shows that climates *were* strongly differentiated and that effective climatic barriers probably existed.

To the south, the *Glossopteris* flora apparently had no limit. *Glossopteris* itself and the coal-forming flora of which it was a part seem to have extended into polar regions. This surprising conclusion apparently must be accepted by both believers in continental stability and believers in continental drift. Coal and apparent *Glossopteris* leaves and conifer needles have been found about 3° from the existing pole, with Antarctica in its present position (Barghoorn 1961). And coal and oil shale were deposited in Tasmania during the Permian and Triassic (Spry and Banks 1962:204, 211, 217, 223–224), when paleomagnetism places the Tasmanian corner of Australia very near the pole (see following paragraph). The Tasmanian coal was deposited in commercial quantities, and some of it contains glacial or interglacial fossil plants, including *Glossopteris*. Whether the continents were in their present positions or whether they were in the positions indicated by paleomagnetism, late Paleozoic coal forests grew in polar or at least subpolar regions in the Southern Hemisphere, and the occurrence of coal on southern continents is therefore not evidence that the continents lay in more favorable latitudes than now. Wegenerians have asked dramatically, "Where was Antarctica when the continent was green?" The answer is that it may have been just where it is now, so far as the evidence of fossil plants and coal is concerned.

Revolutions of climate in the late Paleozoic. Comparison of the paleomagnetic and glacial-botanical records shows something else important. Paleomagnetism places the Tasmanian corner of Australia very close to the South Pole, near or within the Antarctic Circle, in the Carboniferous, Permian, Triassic, and Jurassic (see Fig. 37 on p. 200). During the whole of this time Australia was much nearer the pole than it is now or than it was during the (comparatively light) glaciation of the Pleistocene, if interpretation of the paleomagnetic record is correct. During this long apparent sojourn near the pole, Australia and Tasmania were ice-free and at least partly forested most of the time. But during a limited part of the time ice sheets developed not only on what is now Tasmania but also across the whole southern half of Australia; then, after a while, the ice sheets disappeared again. If the paleomagnetic record is evidence of movement of continents, it is also evidence that Australia moved very little during the time in question and that the glaciation of this and presumably of other southern continents was brought on and ended not by changes of position of the land but by revolutions of climate comparable to the climatic revolution that brought on glaciation on northern continents in the Pleistocene.

This conclusion has a still wider significance. The distribution of climate on the round earth presumably always tends to be symmetric, with temperature decreasing equally toward the North and South Poles, unless some special situation or event interferes with the symmetry. If late Paleozoic glaciation in the Southern Hemisphere was due to a revolution of climate there, the climatic revolution presumably affected the Northern Hemisphere too, and the fact that the northern continents were not glaciated then suggests that they lay farther south than now. The paleomagnetic record indicates that they did so. In the face of these new facts and inferences Brooks's (1949:248, Fig. 29; my Fig. 31) attempt to explain the distribution of late Paleozoic climates by changing the connections (not positions) of continents and by postulating an unsymmetric system of warm and cold ocean currents is no longer satisfactory, if it ever was.

15. Modes of dispersal in the past

Dispersal is a complex process, difficult to understand. What mammals can and cannot do is comparatively simple and obvious, but how even they cross barriers is still argued about, and the dispersal of other animals and of plants is far less understood.

Duality of dispersal. One difficulty is that dispersal is in certain ways a dual phenomenon. This is especially the case among plants, although it is true of animals too. Plants tend to form associations or "communities," which can advance or retreat as wholes. For example, tropical rain forests (as in North Queensland) and also south-temperate *Nothofagus* rain forests (as in Tasmania and southern South America) often form solidly integrated masses of vegetation. These rain-forest communities are glacierlike in their massive coherence and evidently move as wholes at glacial rates.

No one questions that communities can move as wholes. However, some botanists are so impressed by this kind of movement that they think the plants concerned can move only with their communities, or at least that other kinds of dispersal are negligible. Cockayne (1928), for example, in his important book on the vegetation of New Zealand, says (p. 419) that "it is *not* the *species* which move but the *associations* to which they belong." And Burbidge (1960), in her very useful survey of the phytogeography of the Australian Region, says (p. 156), "It is accepted as a principle that plant migration has been by advance and retreat of communities and not, at least in the great majority of cases, by chance arrivals."

That communities move as wholes is obvious in some cases, and within single regions where favorable conditions are continuous this may be the principal mode of plant dispersal for short periods. However, over longer distances, where favorable conditions are not continuous, and during longer periods communities do not behave as units. They were not born as wholes but must have been formed initially by the coming together of plants of various origins. This is evident in principle and is shown by the

different geographic relationships of the members of single communities in many cases. Also, even now, we can see communities
changing by independent movements of single species. An apt
example given by Fleming (1962:94, from Holloway 1954) is the
invasion of podocarp forest by *Nothofagus* in New Zealand. It is a
plain fact that during appreciable periods communities do change
in composition. They are added to and subtracted from, and
sometimes they are decimated and dismembered. This apparently
happened to the Tertiary forests of southern Australia, as described in Chapter 9. Moreover, single species and genera of
plants do disperse for long distances independently, leaving old
associations and forming new ones as they go. The oaks (genus
Quercus) are an appropriate example because they belong to the
same family as the southern beeches (*Nothofagus*) and because
they have dispersed across climatic barriers and have changed
associations as they have done so. If the place of origin or main
radiation of oaks was north of the tropics, different ones have
left their original northern associates and dispersed southward by
two routes, reaching the mountains of New Guinea and meeting
araucarians and *Nothofagus* there, and reaching the northern end
of the Andes and forming associations with palms there (see
p. 63).

Cases like this show that there is indeed a duality in plant
dispersal. Plants often do move as members of associations or
communities. But they also often move for long distances separately, crossing barriers, and leaving old associations and entering new ones. To deny that this can happen, or to deny that it
is important, is self-defeating. The person who denies it defeats
his own efforts to understand the history and significance of plant
distribution. Movement of whole communities of land plants
probably requires continuity of land. To make it an initial principle that plants move only by communities is therefore just one
step from making the former existence of land connections around
the southern end of the world a matter of principle rather than
evidence.

A second sort of duality occurs in dispersal of single species.
In most places, most of the time, most plants and animals probably disperse by slow spreading, or dying back, of continuous
populations. But the same plants and animals can and sometimes do jump wide barriers and establish new populations in

remote places. And the new populations may then behave as if the organisms concerned could not cross even narrow barriers at all! This is illustrated by the Hawaiian fauna (Zimmerman 1948; Darlington 1957:527–529). The Hawaiian Islands are volcanic and coral oceanic islands rising from deep water. They have an endemic flora and fauna derived from a moderate variety of ancestors that apparently reached the islands from several different directions. There is no geologic evidence of even one land bridge to these islands, much less of several bridges from different directions. Also, the composition of the fauna reflects derivation across water rather than across land, and the water gaps have been wide enough to bar all flightless land vertebrates, even lizards. The various land birds, land mollusks, and insects that have reached these islands apparently must have crossed wide ocean gaps. Nevertheless many of them have then radiated, evolving different species on adjacent islands separated by very narrow water gaps, as if the narrow gaps were absolute barriers.

Relative amounts of dispersal. Endemic land birds, land mollusks, and insects, as well as plants, occur on almost every oceanic island suitable to support them, no matter what the distance from other lands. Unless land bridges are postulated to almost every island in the world, it must be supposed that certain terrestrial plants and animals are continually being scattered over the oceans in quantities sufficient to assure establishment of some of them on any habitable island anywhere within a rather short span of geologic time. Distance and the direction of winds and ocean currents affect chances of dispersal, but no island is so remote or so unfavorably placed with respect to winds and currents that it has failed to receive at least a limited terrestrial flora and fauna. A corollary of this fact is that where distances are shorter and winds and currents more favorable the numbers of terrestrial plants and animals that disperse across water gaps are probably relatively enormous. But even a *relatively* enormous amount of dispersal may not be easily detectable. An imaginary example illustrates this. Suppose one propagule (one or more individuals of a given species necessary to establish the species in a new place) from Australia or New Zealand were to reach the forested coastline of southern Chile each year. This would probably be a relatively enormous amount of dispersal, but the chance that a competent observer would be at the right place at the right time to detect the arrival would be infinitesimal.

This example is not only imaginary but also outdated geologically. The chance of propagules of terrestrial plants and animals reaching Chile now, directly from Australia or New Zealand, is probably a minute fraction of the chance of propagules arriving via Antarctica in the past (see pp. 150, 160). The relation of amount of dispersal to distance depends on rate of loss (or rate of death) during dispersal. If the rate of loss is high, a reduction of distance by half may increase the amount of dispersal across a water gap a hundred or a thousand times or more (Darlington 1938a). The theoretical relation of distance to amount of dispersal can, of course, be stated mathematically, but this is not the place to do it. We do not yet know enough about dispersal in the far south to justify treating it with mathematical precision.

The evidence that some terrestrial plants and animals do cross ocean gaps is clear. How they do it is still far from clear, and is one of the outstanding problems of biogeography. Darwin (1859, 1964:356–365) recognized the problem and attacked it by experiments on viability and buoyancy of seeds in ocean water, and ocean transport has been investigated by other persons since then (see especially Guppy 1906; 1917). Aerial transport may be important too. The power of rising air currents and winds to lift and carry small organisms is often underestimated (Darlington 1938a; 1957:17–20). Investigation of what winds really do carry has been begun in the last few decades by use of aerial nets pulled by ships and airplanes in various parts of the world, including recently the antarctic region (Gressitt and Yoshimoto 1964; Gressitt 1964c).

Summary of modes of dispersal. Plants and animals disperse both as members of associations and separately, and both by continuous spreading and by jumping barriers. Any given group of plants or animals may have dispersed in all these ways at different times. Moreover some terrestrial plants and animals evidently disperse across oceans more freely than might be supposed, and they may be expected to do so in relatively enormous quantities where distances are relatively short and winds and ocean currents exceptionally favorable. These conclusions are based on analysis of general situations. How dispersal has occurred in particular cases in the far south is more difficult to decide. However, inferences can be made in some cases, and will be made in the case of *Nothofagus* in the following chapter.

The conclusions summarized above are not shared by all biogeographers. Whether and how terrestrial plants and animals cross ocean barriers has been debated at least since the time of Hooker and Darwin (Chapter 1) and is still debated. A real dilemma underlies the debate. On one hand it is very difficult to see how some kinds of plants and animals can ever get across barriers of salt water. Figs (Corner 1964) and oaks among plants and frogs and *Peripatus* among animals are examples. On the other hand it is very difficult to see how all the places where such plants and animals occur can have been connected by land. This dilemma cannot be resolved by arguments about the dispersal powers of particular plants and animals but only by analysis of the compositions and distributions of whole floras and faunas in critical places. In other words, the irreconcilable *cannot cross salt water* and *cannot have land connections* schools of biogeography should be abandoned in favor of a *let us look at the whole situation and see what did happen* school. This school finds that, as wholes, the floras and faunas of (for example) the West Indies (where endemic figs and oaks and frogs and *Peripatus* occur) as well as of the Hawaiian Islands (where the biota is much more limited) clearly reflect derivation across water gaps. The situations on these and many other islands seem to me to justify the conclusions summarized in the preceding paragraph. For more detailed discussion of this fascinating and important but complex and difficult subject see Matthew (1915, 1939), Darlington (1938a), Zimmerman (1948), and some of the articles in Gressitt (1964d).

16. Geographic history of Nothofagus

The key to the history of terrestrial life in the far south may be *Nothofagus*. Trees of this genus are conspicuous and well known, so that their distribution is well known, and they seem to have no exceptionally great powers of dispersal. No other group of plants or animals is more characteristic of the southern cold-temperate zone, and no other southern group has left so good a record of its distribution in the past. I shall therefore consider the history of this genus with special care to see what can be learned or inferred from it. I am much indebted to Dr. Lucy M. Cranwell for up-to-date information of the pollen record of *Nothofagus*, and especially for letting me see the manuscripts of unpublished papers. However, Dr. Cranwell is not responsible for my conclusions.

The present distribution of *Nothofagus* is described in Chapter 3 and is mapped in Figs. 5 and 20. Geographically, the genus is now confined to the southern tip and western side of southern South America, parts of Tasmania and widely spaced plateaus and mountain ranges in southeastern Australia, New Zealand, and mountains on New Caledonia and New Guinea. Climatically, the genus occurs in areas of moderate to heavy rainfall with oceanic or montane climates (the two are similar) from the coldest part of the south temperate zone in southernmost South America to the tropics or at least subtropics in New Guinea down to 1000 m altitude. The genus seems now to occur in practically all areas that are climatically suitable for it in the Southern Hemisphere except South Africa (if the climate there is suitable) and except some relatively small islands. However, its distribution shows local discontinuities within each main area of occurrence.

Botanists recognize three groups of species of *Nothofagus*. The present distributions of the groups are correlated with climate. Both the *menziesii* and the *fusca* groups now occur in all three south-temperate areas of distribution of the genus: in southern South America, southern Australia–Tasmania, and New Zealand. The *brassii* ("intermediate") group is now confined to New Caledonia and New Guinea, although it too formerly occurred in all three southern cold-temperate land areas.

Although the *menziesii* and *fusca* groups of *Nothofagus* have distributions that are similar in general, the two groups do not usually occur together. In fact, at least in southernmost South America and Australia–Tasmania, any one piece of *Nothofagus* forest usually consists of one species of the genus, and different species have somewhat different climatic requirements or preferences. In southern Chile, for example, the drier eastern edge of the forest consists of *Nothofagus pumilio* (a misnomer!), which is deciduous, while the wetter western edge of the forest consists of *N. betuloides,* which is evergreen. In Australia, *N. cunninghamii* (Tasmania and Victoria) and *mooreii* (scattered mountains in New South Wales and

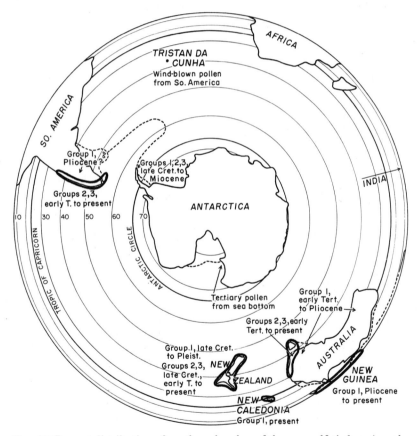

FIG. 20. Present distribution of southern beeches of the genus *Nothofagus* (see also Fig. 5) and summary of the pollen record of the three groups of the genus.

the southern border of Queensland) are evergreen forest trees which often form pure stands, while *N. gunnii* is a deciduous shrub which resembles an alder (superficially, of course) and forms thickets at timber line on mountains in Tasmania. *N. cunninghamii* and *mooreii* belong to the *menziesii* group; *gunnii,* to the *fusca* group.

The insects in different kinds of *Nothofagus* forest differ in some cases, suggesting significant differences in the environment. In southern Chile, for example, the carabid tribe Migadopini is represented in the relatively dry deciduous *Nothofagus* forest along the eastern edge of the forest zone by *Migadops latus,* which is usually the commonest carabid where it occurs at all, but which apparently does not occur in the wetter forest of evergreen *Nothofagus* along the western edge of the forest. A different migadopine, *Antarctonomus complanatus,* occurs in the wet evergreen forest. I found this species in deciduous forest too, at Puerto Williams, but only in places where seepages made the ground wet.

The pollen of *Nothofagus* is produced in enormous quantities and is almost indestructible. The waxy exines or outer parts of the pollen grains can be found and identified in stratified deposits through more than 50,000,000 years of time. Pollen of *Nothofagus* is reasonably easy to recognize and the pollens of the three different groups of the genus are distinguishable. There are, however, several possible sources of error to be remembered in interpreting the pollen record. Errors of identification can be made. Critical identifications should be confirmed by competent specialists, and some specialists think that it is necessary to see actual pollen grains rather than photographs. Detectable quantities of *Nothofagus* pollen can be carried almost unbelievable distances by prevailing winds, for example from South America to Tristan and Gough Islands, about 4500 km or 3000 mi (Hafsten 1951:34–38), and probably even to Antarctica (Holdgate 1964). So the finding of a few pollen grains does not always prove the occurrence of *Nothofagus* close by. Also, when gravel or peat containing *Nothofagus* pollen is eroded and redeposited, the pollen may be redeposited too, so that (say) Eocene pollen may be found in Pliocene strata. And contamination can occur in the laboratories where fossil pollen is examined. Persons examining fossil material often use recent material too, and the herbarium specimens can conceivably shed pollen into the fossil preparations. There is even a chance of contamination of fossil preparations by pollen from living *Nothofagus* trees that are grown in gardens in some places in the Northern Hemisphere, including southern England. Errors from these sources, however, affect only details and not the main, massive record of *Nothofagus* in the Southern Hemisphere. The known record is astonishingly good, although of course incomplete, as fossil records always are. (See Faegri and Iversen 1964 for discussion of the nature, potentialities, and complexities of the fossil pollen record of plants.)

The known pollen record of *Nothofagus* in order of time and completeness, so far as these characteristics are correlated, is as follows (and see Fig. 20).

The record is best, or best known, on New Zealand. Pollen of the *brassii* group appears there first, rather late in the Cretaceous, and is followed by the *menziesii* group, still in the Cretaceous, and by the *fusca* group, in the early Tertiary. All three groups continue in the record on New Zealand until the *late* Pleistocene, when the *brassii* group disappears (Fleming 1962:86). The other two groups still exist on New Zealand, of course. It should be added that fossil remains of other plants, including some pollen, have been found in New Zealand in deposits older than the late Cretaceous (Fleming). That *Nothofagus* has not been found in these earlier deposits suggests an actual arrival in New Zealand late in the Cretaceous.

The record of *Nothofagus* on Antarctica is mostly on the Antarctic Peninsula and adjacent small islands. Moreover, the pollen-bearing deposits are all marine, apparently formed in shallow seas, although some of the plant material in them looks as if it came from land near by. The earliest *Nothofagus* pollen in these deposits is well dated and is from the late Cretaceous, approximately contemporaneous with the first known occurrence of the genus on New Zealand. The latest records from the antarctic deposits are rather doubtfully dated as Miocene. The Eocene as well as the Pliocene is missing from the record, but the probability is that *Nothofagus* forest existed on what is now the Antarctic Peninsula at least from the late Cretaceous to the Miocene. All three groups of the genus were present, and with them occurred a limited assemblage of other plants indicating (I think) a cool-temperate climate. The recovery of a small amount of pollen from the sea bottom near McMurdo Sound suggests the presence of *Nothofagus* on at least the edge of Antarctica proper in the Tertiary but leaves details doubtful (see p. 117, and see again Holdgate 1964).

The actual record of *Nothofagus* in Australia and Tasmania apparently does not begin until the early Tertiary and is more fragmentary than on New Zealand. All three groups of *Nothofagus* occurred in southeastern Australia in the Tertiary, but only the *brassii* group extended west across the continent. Pollen of the other groups has not been found west of Victoria. The *brassii* group disappears from the Australian record at the end of the Pliocene. The other two groups still exist.

In South America, the actual record of *Nothofagus* does not begin until the Upper Eocene and is fragmentary. All three groups are represented, but *brassii*-type pollen has actually been found only well north of the tip of the continent (in the "La Mision profile") and only in relatively recent strata (doubtfully Pliocene). The *brassii* group has of course now disappeared from South America, although the other groups survive there.

In New Guinea, only *brassii*-group pollen has been found and probably not before the Pliocene, although the dating may be doubtful. However, the record in New Guinea may be very incomplete. It hardly justifies a conclusion as to when *Nothofagus* reached the island. In New Caledonia, no fossil *Nothofagus* pollen has yet been found, although the genus occurs there now. Only members of the *brassii* group occur on New Guinea and New Caledonia.

No *Nothofagus* pollen has been found in South Africa (Levyns 1962:239) or on Madagascar and supposed records from India are doubtful, and there is no other evidence that *Nothofagus* has ever existed in these places.

The time and place of origin and direction of spread of *Nothofagus* are not shown by the actual record but can be inferred. The time of origin was probably the Cretaceous. *Nothofagus* is an angiosperm and not an especially primitive one. The main radiation of angiosperms seems to have occurred in the Cretaceous. And *Nothofagus* itself is not known until the late Cretaceous, although by that time it had reached the southern end of the world and had at least begun its dispersal there.

As to place of origin, it has been said that the distribution and pollen record of *Nothofagus* indicate an "entirely southern origin" (Couper 1960:498). The same sort of logic would justify saying that *Fagus* has had an entirely northern origin. But generalities like these get nowhere, for the two genera have evidently had a common ancestor. Cranwell (personal communication) suggests that we shall be near solving the problem of place of origin when we find pollen of *Nothofagus* and *Fagus* together, and we may be on the verge of doing so. Old records of macrofossils and pollen assigned to *Fagus* in the Southern Hemisphere are all wrong or unconfirmed (Cranwell 1964). There are also wrong and unconfirmed records of *Nothofagus* in the north. However, although it is not yet certain, Dr. Cranwell thinks that real *Nothofagus* pollen finally may be turning up in the north, in Siberia and perhaps elsewhere.

A northern origin of *Nothofagus* is suggested also by the facts that the family Fagaceae is mainly northern, and that *Nothofagus* includes apparently potential tropics crossers, members of the *brassii* group, which seem more tolerant of tropical and subtropical climates than *Fagus* is. Either *Fagus* or *Nothofagus* must have crossed the tropics, or a common ancestor must have done so, and the *brassii* group of *Nothofagus* seems reasonably fitted to have done it.

If *Nothofagus* reached the southern end of the world from the north, where and how did the genus cross the tropics? And did it cross just once or did it do so twice, in the Old and the New World separately? There is nothing inherently improbable in the idea that *Nothofagus* might have crossed the tropics twice. The oaks seem to be doing so now. Nevertheless I see no evidence that *Nothofagus* did it and no reason why it need have. If *Nothofagus* did originate in Asia, it was probably never as widely distributed around the north temperate zone as the oaks are. If it had been, we should probably know the fact by now, from the pollen record in the north. *Nothofagus* therefore apparently never had a broad northern base from which to make a double crossing of the tropics. Moreover, although two crossings in opposite hemispheres by a single group of *Nothofagus* might be reasonable, it is hardly reasonable to suppose that the three groups of *Nothofagus* all crossed the tropics twice. It seems much more likely that they dispersed around the Southern Hemisphere. So only one crossing of the tropics seems required.

A reasonably complete and consistent hypothetical history of *Nothofagus* can now be outlined. The genus may have originated in Asia, perhaps primarily in subtropical parts of southeastern Asia, during the Cretaceous. It probably was never widespread or dominant in the Northern Hemisphere. It probably crossed the tropics once, by way of the Indo-Australian Archipelago, in the Cretaceous. The *Nothofagus* of the *brassii* group now on New Guinea may be descendants of the original, Cretaceous tropics crossers. Some botanists think that *Nothofagus* has reached New Guinea much more recently, from the south, and this is possible so far as actual evidence goes. However, the hypothesis of a recent northward movement to New Guinea raises unnecessary complications and difficulties.

An origin in Asia and dispersal southward across the Indo-Australian Archipelago would bring *Nothofagus* into the Southern Hemisphere to Australia or New Zealand or both. A radiation of the genus may have occurred in one or both of these places perhaps in the late Cretaceous. The *menziesii* and *fusca* groups may have originated there, partly by climatic adaptation. Both these groups seem more at home in southern cold-temperate climates than the presumably ancestral *brassii* group is. All three groups apparently then spread much of the way around the world in the Southern Hemisphere.

All this seems straightforward enough, but it leads to a difficult question. How did the three groups of *Nothofagus* effect their dispersals in the south? The first step toward answering this question is to try to decide when the dispersals occurred, and the record, especially on New Zealand, indicates that the three groups spread around the Southern Hemisphere at slightly different times in the late Cretaceous and early Tertiary.

This dating is imprecise and may be wrong, but unless it is very far wrong it seems to rule continental drift out of the question. If drift did occur, *Nothofagus* was probably much too late to be affected by it. If the present distribution of *Nothofagus* fits a Wegenerian pattern of land that may have existed early in the Mesozoic, the fit is presumably accidental and meaningless. No angiosperm genus is likely to date from a time of actual union of southern continents (see p. 167). And the distribution of this genus is closely correlated with present climate (p. 140). If climate determines distribution now, as it seems to do, how can correlation with early Mesozoic geography (if there is a correlation) be anything except accidental?

Whether or not *Nothofagus* may have dispersed around the Southern Hemisphere by means of narrower land connections—isthmian links—is more difficult to decide from evidence of *Nothofagus* itself. The possibility of isthmian connections in the far south will be considered in more detail in the following chapters, and a decision about them (if one can be reached) will be reserved until then. Now, I want to conclude only that *Nothofagus* may have originated in (southern?) Asia in the Cretaceous, crossed the tropics to Australia or New Zealand or both, radiated there, and somehow made a triple dispersal half way around the southern end of the world from there, during the late Cretaceous and early Tertiary. This history is derived from some actual evidence. Nevertheless it is a hypothetical history, incomplete and probably wrong in some details.

The mode of dispersal of *Nothofagus* is suggested by two facts: that existing species often occur in separate stands and in somewhat different habitats, and that the three groups of the genus appear in the record on New Zealand at different times. These facts suggest that the different groups of *Nothofagus* have dispersed separately rather than as members of one community. Separate species of *Nothofagus* can be observed to change their geographic limits by advance and retreat of continuous populations. Whether and how they may also sometimes jump wide gaps, including ocean barriers, is a matter of inference. They have not been observed to do so, but even a relatively enormous amount of dispersal of this sort might not be observable (p. 137). Logs and stumps of *Nothofagus* are known to drift from southern South America to Tasmania (Barber, Dadswell, and Ingle 1959), but they take a long time to do it. Can ocean currents have carried viable *Nothofagus* seeds across narrower ocean gaps, via Antarctica, in the past? I do not know. Dispersal by wind seems to me more likely. Van Steenis (1953:316, 326, 327), discussing primarily the New Guinean forms, says that the seeds of different species of *Nothofagus* differ in size. Some are as large as the "nuts" of northern beeches (*Fagus*) but others are smaller. Also, while the nuts of northern beeches are only sharp-angled, the seeds of *Nothofagus* are "generally winged." Preest (1964) figures the seeds of several New Zealand species and says that they vary from 3 to 10 mm in length and are relatively light, provided with membranous ridges or "vanes," and apparently "well suited for dispersal by wind," although Preest himself doubts if they really are blown long distances.

The importance of *Nothofagus* in southern biogeography is worth re-emphasizing. This exceptionally well-known genus of plants, with its extraordinary fossil (pollen) record, is likely to disclose a geographic history that has been followed by many other plants and by many invertebrate animals in the far south. Comparisons may be extended, tentatively, far back into the past. For example, if *Nothofagus* has been wind-dispersed across southern ocean gaps in the late Cretaceous and Tertiary, *Glossopteris* may have been dispersed in the same way in the late Paleozoic. The seeds of *Glossopteris* were apparently even better adapted for wind dispersal than those of *Nothofagus* (p. 193).

17. Late Cretaceous and Tertiary history of far-southern land and life

The present chapter is intended to bring together what is known and can reasonably be inferred about the geologic, geographic, climatic, and biotic history of far-southern lands from the late Cretaceous to the present. This span of time is relatively recent and relatively well known. It can conveniently be dealt with before the more difficult earlier periods are considered.

General review of late Cretaceous and Tertiary history. All the main pieces of land in the southern cold-temperate zone seem to have been continuously available to life through the late Cretaceous and Tertiary. These lands have not been depopulated and repopulated during the time in question, although complex changes have occurred in their floras and faunas. Even Antarctica, or at least its edges, may have been continuously habitable until formation of the existing ice cap.

All the main pieces of land in the far south seem to have been at approximately their present latitudes during the time in question. Biogeographic (and some paleomagnetic) evidence shows at least that they were south of the tropics. And the southern continents have not been broadly in contact during this time. Even *if* the continents were once in contact and have moved apart, and *if* movement continued into the late Cretaceous, actual separation of the continents apparently must have occurred much earlier (see following chapters). Therefore, *if* land connections existed in the south during the late Cretaceous or later, they were probably isthmian links. And *if* the latter did exist, they probably ran via Antarctica rather than directly between other southern land areas. This is suggested by the geography of the ocean floor and by some geologic evidence. A temporary or partial isthmus between Antarctica and the southern tip of South America is most likely, and might have had important effects on ocean currents, climate, and distribution of plants and animals even if the connection was not itself a dispersal route (see p. 120).

It should be emphasized that, regardless of possible land connections, routes and probabilities of dispersal in the south were presumably very different before Antarctica was ice-capped than they are now. If the edges of Antarctica were habitable in the late Cretaceous and part of the Tertiary, as they probably were, the water gaps that would have to be crossed by terrestrial organisms dispersing in the far south would be much narrower than now, and widths of water gaps are decisive in determining chances of dispersal. If the edges of Antarctica were habitable, chances of dispersal of land plants and animals around the southern end of the world might be hundreds or thousands of times greater than now (see Chapter 15).

Climatically, all the lands now in the southern cold-temperate zone and perhaps also the edges of Antarctica seem to have had generally similar histories from the late Cretaceous through most of the Tertiary. Their climates probably fluctuated but were neither tropical nor glacial until the Pleistocene. Similarity of climate (temperate, and oceanic more than continental in all cases) and of climatic history *must* have tended to cause zonation of life on land in the far south, regardless of the origins and dispersals of the various plants and animals concerned.

The importance of climate can hardly be reiterated too often. Whether or not old land connections have played a part in forming present patterns of distribution of far-southern plants and animals, the patterns are surely due partly to zonation of climate now and in the past. Besides the obvious importance of climate in plant and animal distribution everywhere, there are two special evidences of it in the south temperate zone. One is the close correlation with climate of the distributions of some southern groups now, including the southern beeches (p. 140). And the other is the number of southern groups that either have amphitropical distributions themselves or form amphitropical patterns with related groups in the north temperate zone, showing that climate has been involved in the evolution and dispersal of the groups for a very long time (Chapter 14).

Biotically, all the principal pieces of land in the southern cold-temperate zone have had much in common through the late Cretaceous and Tertiary. They are all known to have had southern conifers since before the Cretaceous and southern beeches (*Nothofagus*) since the late Cretaceous or early Tertiary, and these plants apparently occurred also on the Antarctic Peninsula until the mid-Tertiary and perhaps on parts of the mainland of Antarctica. Both the conifers and the beeches probably formed forests wherever they occurred. Some other plants, including moorland forms, and also various terrestrial invertebrate animals were probably equally widely distributed in the far south, although virtually no fossil record of the animals has been found.

However, the vertebrates of different far-southern land areas are very different now, and probably always have been. Existing terrestrial mammals of Tasmania are monotremes, marsupials, and rodents of the family Muridae; of Tierra del Fuego, Carnivora, one ungulate, and rodents of the families Cricetidae and Ctenomyidae. The land birds of Tasmania and Tierra del Fuego are different and not directly related to each other. Of frogs, Tasmania has leptodactylids and *Hyla;* Tierra del Fuego, perhaps leptodactylids and *Bufo* but not hylids, which do not reach southernmost South America (Darlington 1957:168). Of reptiles, Tasmania has a moderate representation of Australian families of lizards and snakes; Tierra del Fuego, only one or two species of lizards of the American family Iguanidae. No fishes of strictly fresh-water groups reach either Tasmania or Tierra del Fuego. (The small fishes of the salt-tolerant family Galaxiidae that do occur in fresh water on Tasmania and Tierra del Fuego are mentioned on pp. 38, 65.) These existing terrestrial vertebrates of Tasmania and of Tierra del Fuego have been derived independently from the north, from Australia and South America respectively. Probably successions of such vertebrates have invaded the cold southern corners of Australia-Tasmania and South America continually in the past, with replacements in each place but with little or no exchange between the two (Chapter 6). The flightless terrestrial vertebrates of New Zealand are extraordinarily different from those of both Tasmania and Tierra del Fuego and comprise (of native forms) only the Tuatara, one genus of very primitive frogs, and two stocks of lizards, and of course a unique assemblage of flightless birds.

The remarkable differences in present patterns of distribution suggest that the history of terrestrial vertebrates in the far south has been very different from the histories of (say) southern conifers, southern beeches, and some southern invertebrates. In general, although the evidence does not actually prove every detail, it seems probable that special cool-temperate forests and also subantarctic moorlands have been distributed over all the principal habitable pieces of land in the far south during the time under consideration. Many groups of invertebrates associated with the southern forests and moorlands have probably been widely distributed too. But terrestrial vertebrates have not. At first thought this situation suggests dispersal of the plants and invertebrates across ocean gaps too wide for terrestrial vertebrates to cross. This explanation does account for the absence of mammals (except bats) and of most other flightless vertebrates on New Zealand now, and it can account also for the absence of these animals (if they were absent) on Antarctica in the past. However, the lack of relationship between the vertebrate faunas of southern South America and southern Australia-Tasmania requires other, more complex, explanations (Chapter 6).

Role of Antarctica. The place of Antarctica in the history of far-southern life is critical. All that is really known of land life there during the late Cretaceous and Tertiary is that *Nothofagus* and certain other plants occurred on what is now the Antarctic Peninsula at least until the Miocene and may have occurred on parts of the continent proper. However, something more than this can be inferred.

Whether or not the continents have moved, Antarctica has probably been somewhere near the South Pole for a very long time. This is indicated by glaciation in the late Paleozoic and by paleomagnetic data in the Jurassic and is consistent with what is known of the later climatic histories of Antarctica itself and of southern South America, southern Australia, and New Zealand, which lie around Antarctica in different directions. Antarctica is large, a real continent, with a very severe continental climate. Air tends to flow off the land, so that the interior is dry as well as cold, and this has probably been true as long as the continent has been near the South Pole, and the interior of the continent must have been highly seasonal too, with a long period of very short days or continuous darkness in winter. Under these conditions the interior of Antarctica could probably support only a very special, limited flora and fauna. More diverse life might exist on the margins of the continent, and still more on peninsulas and islands off the coast, which might be expected to have a better, more oceanic climate. There is, I think, nothing in the record on Antarctica to contradict these inferences, so far as the late Cretaceous and Tertiary are concerned. During this time terrestrial plants and animals may have occurred mainly on the margins of the land and especially on the peninsula and outlying islands.

Setting aside the preceding conclusions for a moment, what roles might Antarctica conceivably have played in the history of life in the far south during the late Cretaceous and Tertiary? There seem to be three overlapping possibilities.

First, if Antarctica was isolated geographically as it is now, if only parts of the coast were forested, and if the climate was such as to limit plant and animal life rather strictly, it might have been a dead end. A limited variety of cold-tolerant plants and animals might have reached the continent and survived for a while and then died out, with little or no return dispersal out of Antarctica.

Second, Antarctica might have been a stepping stone in dispersal, especially if the edges of the continent were continuously forested and could act as dispersal paths for appropriate plants and animals.

And third, Antarctica might have been a real center of evolution, where important groups of plants and animals evolved and diversified and from which they spread.

This third possibility, that Antarctica might have been a center of evolution and dispersal, is favored by some botanists and entomologists but seems improbable to me. Any possibility that Antarctica was the center of a pre-Cretaceous radiation of angiosperms seems to be nullified by the fossil record (p. 117). The main, interior part of Antarctica is not likely to have been very hospitable in later geologic periods (if it ever was), so that areas large enough for evolution of new, dominant groups are not likely to have been available. And distributions of plants and animals in the southern cold-temperate zone now do not seem to reflect spreading from an antarctic evolution-dispersal center. The southern beeches, so important in other ways in southern biogeography, illustrate this point too.

I have elsewhere (1957:484–487, Fig. 57) stressed the difference between relict and immigrant patterns in the distribution of life on islands. Relict patterns are formed by the partial extinction of old, widely distributed groups of plants or animals. Different forms tend to survive in different places, and the patterns that result are characterized by irregularities and discontinuities. Immigrant patterns, on the other hand, are formed by spread across series of islands. These patterns are characterized by continuity, by occurrence of continuous series of related forms on successive islands.

The three main pieces of land in the southern cold-temperate zone are biotic islands, and it can be asked whether the distributions of significant plants and animals on them form relict or immigrant patterns (Fig. 21). If the floras and faunas that exist on these small, widely separated pieces of land are relicts of a larger

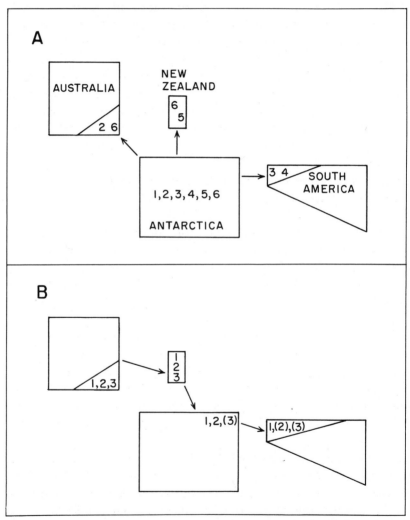

FIG. 21. Diagram comparing hypothetical relict (*A*) and immigrant (*B*) patterns of distribution in the southern cold-temperate zone and Antarctica.

biota that was widespread on Antarctica in the past, we should expect to find *different* fractions of the old flora and fauna surviving in different places. This is what usually does happen when different small pieces of land are separated from one continent. For example, Sumatra, Java, and Borneo have all been separated from southeastern Asia recently, and differential extinction has already begun to make a relict pattern in the distribution of mammals on the islands (Darlington 1957:489–490). But *Nothofagus* in the southern cold-temperate zone does not show a pattern like this. All three groups of the genus occur or have occurred on all three main pieces of land in the far south, and all occurred on the Antarctic Peninsula too in the past. The regularity of this pattern strongly suggests that *Nothofagus* has followed a path around the world in the far south and has not survived in three separate relict areas after evolution, diversification, and extinction on the Antarctic Continent. Of course the coast of Antarctica may have been a dispersal path even if that continent was not the dispersal center.

The nonexistence of land birds that might be relics of a former antarctic fauna is another indication that Antarctica has not been an important evolutionary center, at least not in the Tertiary. That special groups of land birds have not evolved in the relatively small cold-temperate areas on the southern corners of South America and Australia is understandable, but if a large part of Antarctica was habitable, land birds might be expected to have got there and to have evolved special groups adapted to the antarctic climate (as ptarmigan are to the arctic, for example), and some of these birds might survive in Tierra del Fuego or Tasmania. No such survivors are recognizable. The peculiar land birds on New Zealand appear to be products of evolution on an isolated island rather than relics of an antarctic avifauna.

Some groups of plants and animals that are much differentiated in different places in the southern cold-temperate zone, for example carabid beetles of the tribe Migadopini (Chapter 3), might be relics of radiations from Antarctica, but they might equally well be products of old dispersals around the southern end of the world followed by local geographic differentiations. These groups are noncommittal as to the role of Antarctica in dispersal in the past.

My conclusion about the role of Antarctica since the late Cretaceous, based partly on inference about its probable climatic history, partly on the distribution of significant plants and animals on lands around it now, and partly on lack of evidence to the contrary, is that the Antarctic Continent has not been an important center of evolution and dispersal but that many far-southern plants and animals may have dispersed along its edges.

Late Cretaceous and Tertiary land connections? Although broad contacts between southern continents cannot have existed as late as the late Cretaceous and Tertiary, isthmian connections are at least possible, and the evidence for or against them must be considered carefully. Biogeography supplies pertinent evidence from at least three sources.

First, the record of mammals does not require any southern land connections and limits the ones that can have occurred. No complete, usable land connection between South America and Australia can have existed later than the end of Cretaceous (Chapter 7). A hypothetical Tertiary bridge that would carry south-temperate forest, subantarctic moorland, and associated invertebrates but that would not carry mammals is not worth serious consideration. Does the pollen record of terrestrial plants indicate that dispersal stopped abruptly late in the Cretaceous, as if land connections were broken then? I think not. The mammalian record might allow (but does *not* suggest) successive connections even in the Tertiary, first from Australia to Antarctica and later, after the first connection was broken, from Antarctica to South America. This would permit dispersal only in one direction and would account for the failure of South American mammals to reach Australia. This is a logical possibility, but I do not think it is supported by any real evidence. A possible Tertiary connection between Antarctica and South America is further discussed in the fourth following paragraph.

Second, the diversity of plants and animals that show relationships between southern South America, southern Australia–Tasmania, and New Zealand does not suggest dispersal across narrow isthmian connections. The latter would probably be selective, allowing dispersal only of groups adapted to a narrow range of climatic and ecologic conditions. Land connections do select in this way. For example, the connection that existed between eastern Siberia and Alaska at times during the Pleistocene allowed passage mainly of plants and animals adapted to the tundra (Rausch

1964). But this kind of selection does not seem to have occurred in the far south. The plants concerned range from forest trees to moorland cushion plants, and they live in climates that range from warm-temperate to subantarctic.

If selection has occurred during dispersal in the far south, it has been in favor of water-demanding forms. The most characteristic southern cold-temperate plants and animals are concentrated in the areas of heavy rainfall, and some of the insects have aquatic larvae. Water-demanding groups might have been selected by passage across a wet southern land connection. But their distribution may also be explained by the distribution of climates. The small wet areas in the southern cold-temperate zone are and probably long have been isolated by barriers of comparatively dry country (Chapter 13), so that water-demanding organisms in the far south are cut off from the rest of the world and are likely to survive for relatively long periods regardless of how they originally dispersed.

The third source of evidence about southern isthmian links comes from New Zealand (Chapter 10). New Zealand shares a large proportion of the groups of terrestrial plants and animals that are common to the southern cold-temperate parts of South America and Australia. The record of the terrestrial animals is too poor to be significant, but the relatively good fossil record of plants on New Zealand does not show that the plants arrived at one time or from one direction. Rather, they seem to have trickled into New Zealand from time to time and from several directions, as they would be expected to do if they came across water rather than across land. Making and breaking of land connections in other parts of the world have sometimes been dramatically recorded in the fossil record. No such events appear in the record on New Zealand. Moreover, the composition of the existing New Zealand fauna, especially the absence of native mammals except bats and the fewness and extreme differentiation of other flightless land vertebrates, virtually proves that New Zealand has not been connected with any other habitable land during the late Cretaceous and Tertiary. If New Zealand has received its flora and fauna across water, isthmian links elsewhere in the far south are unnecessary and unlikely.

Although the evidence summarized above seems strongly against southern isthmian links in general, the possibility of one between southern South America and Antarctica is not ruled out. The geologic activity of the Andean zone of mountain formation, which apparently extends to western Antarctica, is consistent with formation of a land connection at times during the Tertiary. The existence of such a connection, if it did exist, would probably have changed southern climates and affected distribution of southern life both on land and in the sea (pp. 120–121). The fossil record of shallow-water marine animals may show, eventually, whether or not this occurred.

Dispersal around the southern end of the world. I think that we are forced by the evidence to conclude that, so far as late Cretaceous and Tertiary times are concerned, Antarctica has probably not been an important evolution-dispersal center and that the southern continents have probably not been either in contact or connected by any general system of isthmian links. Nevertheless many terrestrial plants and invertebrates seem to have dispered around the southern end of the world during this time. If this is correct, dispersal must have been across the water gaps, somehow or other. I do not want to minimize the difficulties of this conclusion but want to begin with the statement just made, that we are forced to it, and then discuss some possibilities.

Did floating ice carry terrestrial plants and animals around the southern end of the world during the Cretaceous and Tertiary? Probably not. The geologic record shows little glaciation on land in most of the places during most of the time under consideration. Ice can therefore probably be ruled out as a general means of dispersal most of the time, although ice can carry land organisms in various ways, and although it may have been important in the Pleistocene.

Birds too can carry some land plants and invertebrates in various ways, but birds can hardly have been the principal agents of dispersal of terrestrial life in the far south. Some of the most widely distributed groups of southern organisms, for example the southern beeches, do not seem to be adapted for dispersal by birds. And individual birds probably do not often cross far-southern water gaps from one piece of land to another. However, the possible role of birds as dispersal agents needs further investigation.

There remain, I think, only ocean currents and winds as agents adequate to produce the observed results. I have already noted (Chapter 15) that some terrestrial plants and animals somehow cross the widest ocean gaps and populate the most remote oceanic islands even where currents and winds do not seem especially favorable, and that the amount of dispersal where distances are shorter and where currents and winds are more favorable should be relatively enormous. Habitable lands in the far south are very far apart now, but distances between them were much shorter when parts of Antarctica were habitable. Tremendous winds and ocean currents circle the southern end of the world now and have probably always done so, unless the current pattern was changed by a land barrier. And a barrier, for example a land connection between Antarctica and southern South America, might increase rather than decrease chances of drift dispersal from one far-southern land to another.

Chances of wind dispersal are harder to assess but are probably better than might at first be supposed. The power of air currents to lift and carry small objects is grossly underestimated by most persons. This is one subject where man's intuitive judgment is contradicted by mathematics (Darlington 1957:17–20). Strong winds do circle the Southern Hemisphere at the right latitudes. And investigations are in progress to discover what these winds do carry now. Gressitt and Yoshimoto (1964:286) think that storm winds may be most effective in dispersal, but sampling them is difficult.

Although I am driven by failure of other reasonable possibilities to conclude that terrestrial plants and animals have probably been dispersed around the southern end of the world by ocean currents and winds, I do not pretend to know how any one kind of organism has crossed any one water gap. This problem should be studied experimentally. An obvious candidate for such a study is *Nothofagus* (Chapter 16).

Summary. In summary of the history of southern land and life since the late Cretaceous: no great changes in arrangement or connections of land seem indicated except possibly the making and breaking of an isthmian link between Antarctica and southern South America. No great changes of climate seem to have occurred in the southern cold-temperate zone until the Pleistocene, although fluctuations between warm-temperate and cool-temperate periods perhaps did occur. Whether the interior of Antarctica was habitable during the late Cretaceous or Tertiary is doubtful, although the coasts of the continent and the Antarctic Peninsula (or Archipelago) may have been forested until the late Tertiary.

The history of terrestrial life in the far south during the late Cretaceous and Tertiary has evidently been complex, and I have necessarily oversimplified it. Various plants and animals have probably been arriving from the north continually, the southward movements in some cases being products of world-wide patterns of evolution and dispersal. Some groups have probably reached the south just once, by one route; others, several times, by several routes, in both hemispheres. Whatever their origin, some groups of plants and invertebrates (but not terrestrial vertebrates) have become specially adapted and confined to the southern cold-temperate zone, and some of them seem to have spread around the world there, apparently across the ocean gaps. The result has been the accumulation of a characteristic far-southern biota, much of it now occurring on all three main pieces of land in the southern cold-temperate zone and probably formerly on the edges of Antarctica too, and including especially plants characteristic of wet forest and wet moorland and invertebrate animals living in these habitats. Finally, late Tertiary and Pleistocene climatic changes decimated biotas everywhere in the far south, changed their distributions locally, and probably stopped or virtually stopped dispersal across the water gaps by making Antarctica wholly uninhabitable and unavailable as a stepping stone in dispersal.

18. Permo-Carboniferous and Mesozoic history (1): Africa and South America

The history of far-southern land and life before the late Creta-ceous is more difficult to fathom than the later history treated in the preceding chapter. To trace it correctly from southern evidence alone may be impossible. A different method is necessary. The method is to unite the history of the southern end of the world with that of the world as a whole. And the unifying question is, was the arrangement of all the continents on the world during and before the Mesozoic essentially the same as now or very different? Actually, the central, exciting question now is, have or have not the continents "drifted"? The present chapter summarizes the history of this subject and treats the special case of Africa and South America, for these two continents yield important evidence of the probability of drift and the possible time of it.

Brief history of ideas about continental drift. The idea that continents may have moved has caused arguments and raised emotions certainly since the time of Charles Darwin and perhaps longer. The arguments have sometimes been marked by ignorance, bias, closed minds, overenthusiasm, and blindness to evidence occasionally amounting to suppression of evidence. I do not propose to accuse any particular persons of these faults, but shall try to keep reasonably free of them myself. I know and shall admit my own ignorance of some pertinent subjects, and I know I am biased to some extent. I am biased against moving continents or making other fundamental changes in the world unless real evidence requires it.

That the continents may have moved is an old idea. Francis Bacon, in *Novum Organum* (1620; I have consulted only an 1859 English edition), compared the shapes of Africa and South America and called their similarities "a coincidence which is not accidental," but I doubt if he meant to suggest that the continents had been united. Beginning in the late 1850's, if not before, the idea of continental drift was apparently widely discussed and occasionally written about. Antonio Snider (1859, facing p. 314; my Fig. 22) actually published a map that is something like Wegener's later map of the world before continental separation. Some additional details of the history of the subject are given by Wegener himself in later editions of his book (e.g. 1937:2–4), but Wegener's historical sketch is very incomplete. A more comprehensive history of pre-Wegenerian ideas of "drift" would be worth the while of some historically minded geologist to compile.

Charles Darwin, who knew both the geologic and the biologic aspects of the question, evidently thought carefully about movement of continents, although he summarized his conclusions in few words. He said (1859, 1964:357):

> It seems to me that we have abundant evidence of great oscillations of level in our continents; but not of such vast changes in their position and extension, as to have united them within the recent period.

Fig. 22. Antonio Snider's (1859) suggested arrangement of continents before separation.

The significant word here is "position." Darwin's "recent period" was not our Recent but was evidently intended to cover the whole time of evolution and dispersal of existing groups of plants and animals. Darwin said also (p. 343) that

as far as we can see, where our oceans now extend they have for an enormous period extended, and where our oscillating continents now stand they have stood ever since the Silurian epoch; but . . . long before that period, the world may have presented a wholly different aspect.

Whether or not Darwin was correct in detail, he at least made a deliberate effort to restrict his conclusions to what the evidence actually seemed to show, and he admitted the limits of the evidence. How many Wegenerians and anti-Wegenerians have followed his example?

Somewhat later one of Charles Darwin's sons, George Darwin, and Osmond Fisher suggested that the moon originally broke away from the earth, from what is now the Pacific side, and took much of the earth's crust with it, and Fisher (1882) suggested that what was left of the crust broke into pieces which were pulled apart as the earth readjusted its shape and which became the existing continents. This happened, if by any chance it did happen, early in the earth's history, long before life existed on earth. This old theory might explain the matching shapes of continents but does not explain other phenomena that are part of the current theory of continental drift.

Wegener presented his more detailed hypothesis of the origin of continents and oceans in journal articles in 1912 and then in his famous book in 1915. He thought that all the continents were united in a single land mass, or "Pangaea," late in the Paleozoic and that they have separated and drifted to their present positions since then. His ideas have been modified by many later authors. The volume *Continental Drift* edited by S. K. Runcorn (1962) gives some idea of how the subject stands now, and a few other recent references are given in the following pages.

The mechanics of drift—of continental displacement—will not be debated here. I am not competent to debate them usefully. I shall therefore say only that geophysicists are still arguing among themselves what forces might displace continents, if any can, and that convection in the earth's mantle is probably the most favored mechanism now (Runcorn, pp. 29ff). What I want to try to decide is whether the continents have in fact been displaced. The evidence bearing on this question now comes from several independent sources, some of them unknown in Wegener's time. Two questions in the general hypothesis of continental drift are critical and can perhaps be answered now. First, what has been the history of Africa and South America? And second, what were the positions of the southern continents when they were glaciated late in the Paleozoic?

Africa and South America. Historically, continental drift was first suggested to explain the matching shapes of continents, especially the fit between the edges of Africa and South America. The fitting together of the edges of these two continents has been figured many times, most accurately by Carey (1958:223, Fig. 21; reproduced by Dicke 1962:659; my Fig. 23). The fit is very good, especially at the 2000-m isobath, and is strong evidence that these two continents have been united.

The matching of shapes of Africa and South America is said to be reinforced by the matching of successive strata and of other geologic structures at corresponding points on the continents' edges (du Toit 1927; Caster 1952), but I am not able to judge how good the agreement is or what it means (but see p. 177). Two explanations of similarity of stratification seem possible. One is that Africa and South America were united when the strata were deposited. The other is that corresponding strata on the two continents were laid down in similar climates, which might mean only that the two areas of deposition were then at nearly (not necessarily exactly) the same latitudes, as they are now. This is one of the several situations in which climate rather than movement of continents may explain situations stressed by Wegenerians.

The significance of the matching shapes of Africa and South America is greatly increased by new knowledge of the Mid-Atlantic Ridge. This ridge on the ocean bottom conforms in detail to the shapes of the continental margins, so that we now have three sinuous lines instead of two that agree remarkably. The agreement in shape of these three lines can hardly be accidental, although Wegenerian "drift" is not necessarily the only explanation of it. Present knowledge of the ocean bottom suggests further that the Mid-Atlantic Ridge may mark an original line of separation of the continents, and that the ocean floor may still be

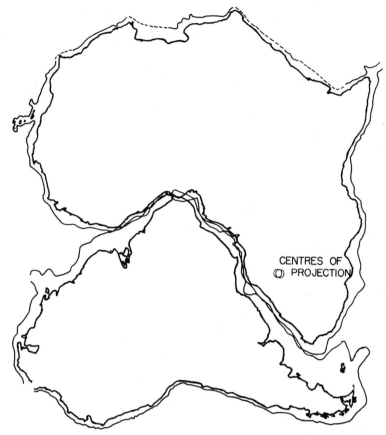

CENTRES OF
(O) PROJECTION

FIG. 23. Fitting of the edges of Africa and South America at the 2000-m isobath (from Carey 1958, Fig. 21, redrawn).

spreading there, gradually increasing the distance between Africa and South America. Wilson (1963a) presents what seems to be strong evidence of this, but some geologists think he overstates the case, and Wilson himself ends by saying that his presentation is still "highly speculative."

Although I am not competent to discuss the technical aspects of Wilson's presentation, I have a layman's comment to make about it. The situation in the Atlantic, considered alone, does seem to show that South America has been separated from Africa and pushed westward by convection. But on the other side of South America, in the Pacific, is another submarine ridge comparable to the one in the Atlantic (Wilson, map on pp. 90–91; Heezen and Ewing 1963). Oceanic islands seem to increase in age with distance from the ridge in the Pacific (Wilson, p. 98) just as they do in the Atlantic. And the oldest oceanic islands in the Pacific seem to be no older than the oldest ones in the Atlantic. No strictly oceanic islands or remains of such islands (seamounts and the like) in any ocean have been found that seem to be older than the Cretaceous (Wilson, pp. 95–96; Menard and Ladd 1963: 382). So far as the geologic structure of ocean bottoms and the ages of islands are concerned, South America may as well be moving toward Africa as away from it. Perhaps it is not moving in either direction. Convection currents from both sides may be diving downward under South America rather than moving the land.

What all this means, I think (of course I am still speaking as a layman), is that the ocean bottoms are yielding evidence of existence of convection currents that might move these and other continents but not, thus far, evidence that specific continents have been moved in specific directions at specific times. The only real geographic-geologic evidence that Africa and South America have been in contact is still (I think) the old, obvious, but very significant fact of their matching shapes. The ages of intervening islands may show the latest time that these two continents can have separated (if they were united), which was probably not later than the end of the Jurassic, but the ages of the islands do not show how much earlier separation may have occurred.

Paleomagnetic determinations do not show east-west positions of continents and therefore do not show how far apart Africa and South America may have been at different times in the past. However, paleomagnetic data do seem to show that the two continents had rotated *independently* before the Jurassic (Creer 1958). This suggests separation of the continents much earlier than supposed by Wilson, very early in the Mesozoic, or before, and recent paleomagnetic calculations hint that the separation may actually have occurred just before the end of the Paleozoic (but see Postscript).

If Africa and South America were still united at the time of late Paleozoic glaciation, the pattern of movement of the ice sheets may prove it, but details of the glaciation, especially of South America, are probably not yet well enough known to justify a conclusion about this, so far as I can judge from Schwarzbach (1963:147–148) and other sources.

African–South American biogeography. Biogeographic evidence of union and of time of separation of Africa and South America might have been decisive but is actually negative or ambiguous. The angiosperms exemplify this fact. They are dominant plants on land. They have dispersed over the whole world—over all habitable continents and to all habitable islands no matter how remote—overwhelmingly and complexly, and also rather recently, geologically speaking. No undoubted angiosperms are known fossil before the Cretaceous. Many important families and genera of them are confined to the tropics or nearly so, and phytogeographers are still debating the routes of dispersal of the tropical groups. Details are beside the point now. The main point is that, if Africa and South America were joined together when the angiosperms dispersed, probably in the Cretaceous, these plants should show a preponderance of relationships between the floras of Africa and South America. However, angiosperms do not show this. Van Steenis (1962:247–253, 314) finds as many plant taxa, most of them angiosperms, shared by the Indo-Malaysian region and tropical America as by Africa and tropical America.

The angiosperms have an additional significance. Most families of trees that now dominate tropical rain forests are angiosperms, and if the rain-forest trees dispersed without benefit of contact between Africa and South America, the insects and other animals in the rain forests may have done so too.

Some existing groups of plants and animals, including conifers and some insects, did make initial dispersals within the time when a connection might have existed between Africa and South America. However, the distributions of these groups do not seem to reflect dispersal across such a connection. Other routes seem to have existed even for tropical organisms, including angiosperms (above), and the older groups may have redispersed long after their time of origin. The conifers have done so. Although some now-extinct conifers were widely distributed in the late Paleozoic, existing families are known only since the Jurassic or, in a few cases, the Triassic (Florin 1963:176). Araucarians, which are among the oldest existing conifers, are known only since the late Triassic and were widely distributed on northern continents then (Florin, p. 178, Fig. 14; my Fig. 32), although they are now confined to parts of tropical and subtropical South America and the Australian region including New Guinea. Neither the fossil record nor the present distribution of these plants suggests a closer connection between Africa and South America than between other tropical regions. Podocarps, another group of conifers, are known only since the Jurassic and are still important forest trees in some tropical and south-temperate regions. Their present distribution and fossil record (Florin, Figs. 16–25; my Fig. 33) indicate a complex dispersal in the Southern Hemisphere but suggest less exchange between Africa and South America than between other southern regions.

If existing dominant groups of tropical forest trees, including angiosperms and podocarps, had dispersed across an African–South American connection, their distributions ought to show it unmistakably. They do not show it. Therefore I think that these trees, which dominate existing tropical rain forests, have probably dispersed by other routes. And I think also, and repeat for emphasis, that insects and other animals in the rain forests have or may have dispersed by other routes too, and that their distributions are not evidence that Africa and South America have been connected.

Fresh-water fishes of Africa and South America. Existing fresh-water fishes (considered briefly in Chapter 7) have an extraordinary distribution that may preserve some details of late Mesozoic (Cretaceous) geography. I have discussed their geographic history at length elsewhere (1957:88–101). It is enough now to say—it is an oversimplification—that the fresh-water fishes of South America are almost all related to African fishes and that this fact has been supposed to be evidence of former union of the continents, but that analysis of the situation changes its significance. The South American fish fauna has evolved from very few ancestors. Although the latter may have come from Africa, they do not represent an old fraction of the African fish fauna, as would be expected if they reached South America before a separation of the continents. On the contrary, the South American fishes seem to be derived from ancestors of various evolutionary levels and various ages that probably reached South America at various times, probably with difficulty, and most likely (I think now, after seeing Harrington's paleogeographic maps of South America) through the sea, not necessarily directly across the Atlantic but perhaps along coast lines and across narrow ocean gaps. In any case, most of these fishes probably dispersed later than the latest time of continental contact that can reasonably be postulated, and they are therefore not evidence either for or against former union of South America with Africa. No strictly fresh-water group of fishes of Africa or South America has reached Australia, and this is hardly consistent with Australia's having been joined to either continent since the origin of the fishes in question.

Fossil reptiles of Africa and South America. Cretaceous reptiles do not show especially close relationships between the faunas of Africa and South America (see Colbert 1952; Romer 1952). Jurassic reptiles are relatively poorly known. However, the mid-Triassic fossil reptiles of South Africa and South America are so similar as to suggest the possibility of a direct land connection at that time (Romer). But this is only a possibility. The evidence is very incomplete. Too many critical questions about the distribution of reptiles in the Triassic are still unanswered (Darlington 1957:611–612).

The distribution of reptiles even before the Mesozoic may give hints of the state of the world. Olson (1955) thinks that reptiles were able to spread between the Old and the New World not long before the Permian, but that early in the Permian exchange stopped and unrelated groups of reptiles evolved separate but parallel adaptive types in the Old and New Worlds. If this is correct (Olson says his interpretations are based on "certain assumptions"), it suggests that Africa and South America separated before the end of the Paleozoic, if they ever were united.

Mesosaurs are a special case. They have been found fossil in South Africa and South America about the beginning of the Permian, and are not known anywhere else at any time. They were small (about a yard long), active, aquatic reptiles, which lived in fresh and perhaps also in salt water (Caster 1952:130). Their ability to cross ocean barriers has been doubted, but, comparing them with some crocodiles and some fresh-water turtles, I think they may have been able to enter the sea and disperse through it, and that no limit can be set to the widths of sea they may have crossed. Their distribution is not good evidence of continuity of land.

These examples are just fragments from a very large, largely controversial body of biogeographic evidence about the history of Africa and South America. In sum, this evidence seems to me to show no definite time of separation of Africa and South America and no clear indication that the continents were ever united at all. If these continents were united, they apparently separated too long ago to leave any recognizable traces of the union in the distribution of existing or relatively recent fossil plants and animals. The distribution of reptiles early in the Mesozoic may be consistent with union of Africa and South America then, but hardly requires it.

Summary of history of Africa and South America. The fitting of their shapes strongly suggests that Africa and South America have been united. Convection currents in the earth's mantle may have separated the continents and moved them apart. If these continents were united, the latest time of separation allowed by geologic dating (ages of islands) is probably the end of the Jurassic, and other evidence seems consistent with separation during the Triassic, or earlier.

19. Permo-Carboniferous and Mesozoic history (2): nonpaleomagnetic evidence

Most Wegenerians have been possessed and, I think, blinded by a fixed idea: that, before drift, the existing continents must have been united in a great supercontinent, an all-embracing Pangaea or at least a southern Gondwanaland (Figs. 24–28). This idea is inherited from Wegener himself (and from his predecessors) and seems to find support in the apparent former union of Africa and South America. However, union of Africa and South America need not require or imply that other continents were united too. Actual evidence of union should be looked for in each case, and evidence that continents were united should be carefully distinguished from evidence that continents have moved. Critical evidence comes especially from late Paleozoic time, from the late Carboniferous and early Permian periods (the Permo-Carboniferous, see Frontispiece) when the southern continents were heavily glaciated. The situation in the Southern Hemisphere during this small segment of geologic time is therefore especially important.

Evidence for or against actual union of southern continents might be derived from fitting of the continents' shapes, from continuity of geosynclines and other geologic structures, from distribution and directions of movement of Permo-Carboniferous ice, from distribution of floras during and after the time of glaciation, and from paleomagnetism. A great deal of evidence does exist, but it has been thoroughly misinterpreted. Much of it strongly suggests that most continents lay farther south on the world in the Permo-Carboniferous than they do now, but no evidence clearly indicates a general union of southern continents.

Fitting of shapes. A general fitting together of continents has often been attempted. Snider, in 1859 (see my Fig. 22), and Baker, in 1911 (see du Toit 1937:14), placed all the continents together and fitted them fairly well, excepting Antarctica. Wegener's hypothetical arrangement of continents "before translation" (that is, before drift) is well known. Du Toit's (1937) arrangement is diagrammed in my Fig. 24. The geometric fitting of continental shapes in this diagram seems neither better nor worse than in various other arrangements. This one is supposed to bring together parts of what du Toit called the Samfrau Geosyncline, but I doubt the geologic necessity and significance of this (see p. 177).

Among more recent Wegenerians (if they do not object to this name), Carey (1958) first emphasizes the remarkable fit of Africa and South America and then offers a restoration of "Pangaea" (his p. 277, Fig. 39b; my Fig. 25) in which the other southern continents fit very badly. Carey (p. 316) says also that he has tried for years to fit all the continents together in an acceptable Pangaea and that he cannot do it on the earth at its present size. He therefore suggests an expanding earth, smaller when (if) the continents were all joined together than it is now. I cannot judge the technical arguments, but (speaking as a layman) I should think that the present almost exact matching of the edges

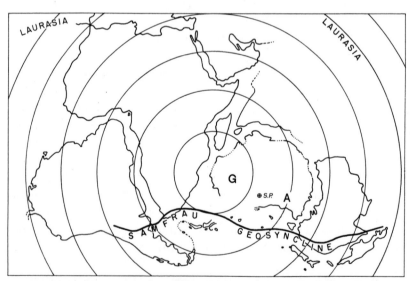

FIG. 24. Du Toit's "reassembly of Gondwana during the [late] Palaeozoic Era" (from du Toit 1937, Fig. 7, redrawn and simplified): *G*, approximate glacial pole (center of the glaciated areas); *A*, approximate Australian paleomagnetic pole (mean of several late Paleozoic pole positions from Runcorn 1962, Fig. 24 [my Fig. 37], and Creer 1964, Fig. 7 [my Fig. 27]). Du Toit thought that the spaces between continents were then mostly land.

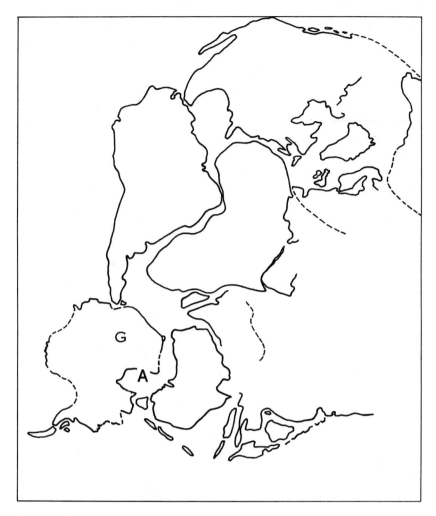

Fɪɢ. 25. Carey's "restoration of Pangaea" (from Carey 1958, Fig. 39b redrawn and simplified); *A, G* as in Fig. 24.

of Africa and South America demonstrated by Carey himself implies that when (if) these two continents were joined the earth was almost exactly the same size as now. I do not think Carey mentions this argument against his theory of an expanding earth.

In King's (1961:310, Fig. 1; my Fig. 26) suggested arrangement of continents for the late Carboniferous, the fit appears fair, but this is partly because of the small scale of the diagram. I should perhaps add now that the pattern of ice movement indicated in the diagram is largely wrong or doubtful (see pp. 182–183).

Creer's (1964a, Fig. 7; my Fig. 27) diagram of "Gondwanaland" in the late Paleozoic is almost the same as du Toit's and is further discussed on (my) pages 205–206.

In Wilson's (1963a, b) suggested arrangement of continents in the mid-Mesozoic (my Fig. 28), which supposedly brings the continents together partly according to guidelines (ridges) on the ocean floor (see fifth following paragraph), the fit is poor in many details.

Runcorn (1962:33, Fig. 24; my Fig. 29), in his "attempted reconstruction of the southern hemisphere for mid-Mesozoic time" based on paleomagnetism, wisely does not even try to fit the continents together exactly. (And note that paleomagnetic evidence does not justify even the close grouping of continents that Runcorn does suggest, as I shall explain in Chapter 20.)

This sampling of Wegenerian plane geometry—of attempts to fit the shapes of the southern continents together—may be concluded by stressing the contrast between the precise fit of the edges of Africa and South America and the failure to fit of the other continents. The failure is emphasized by the variety of ways in which Wegenerians have tried to put the other continents together, unsuccessfully. The contrast between the very good fit of Africa and South America and the failure to fit of

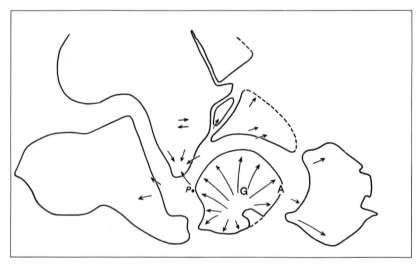

FIG. 26. King's version of "Gondwanaland" in the late Carboniferous (from King 1961, Fig. 1, redrawn and simplified); *A, G* as in Fig. 24. Although the continents are shown separated, King (p. 310) thinks they actually formed "a single land mass . . . during Palaeozoic and early Mesozoic time."

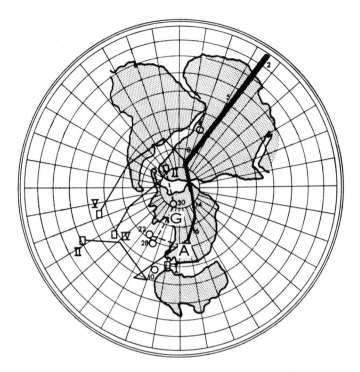

FIG. 27. Creer's "reconstruction of the continents for the Upper Palaeozoic" (from Creer 1964, Fig. 7; copyright *Nature*); A, G as in Fig. 24.

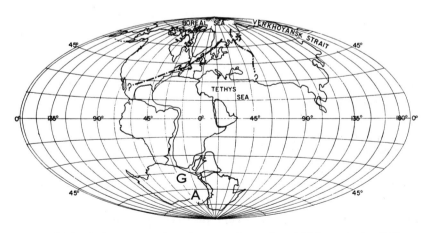

FIG. 28. Wilson's "reconstruction of the continents in mid-Mesozoic time" (from Wilson 1963b, Fig. 6; copyright *Nature*); A, G as in Fig. 24.

Fɪɢ. 29. Runcorn's "attempted reconstruction of the southern hemi-sphere for mid-Mesozoic time" (from Runcorn 1962, Fig. 24; copy-right Academic Press); *A, G* as in Fig. 24. Note that the paleomag-netic data on which this reconstruction is based indicate the latitude and orientation of each continent separately and do not require that the continents be so closely grouped on one side of the pole.

other southern continents strongly suggests that the others were not united or that, if they were, they separated long before Africa and South America did.

Geologic evidence. The evidence of Permo-Carboniferous glaciation (following pages) and of paleomagnetism (Chapter 20) will be discussed separately. First, however, some other geologic phe-nomena should be considered briefly.

Du Toit (1937) and other Wegenerians have placed great stress on supposed continuity of a synclinal system, the "Samfrau Geo-syncline" (Fig. 24), connecting southeastern South America, the southern tip of Africa, western Antarctica, and eastern Australia, and supposedly indicating that these pieces of land were once aligned and in contact. Doumani and Long (1962) have revived the Samfrau reconstruction and overstated the case for it and for southern continental contacts in general. Two serious criticisms can be made of the Samfrau hypothesis. First, part of the supposed geosyncline may not exist; Harrington (1962) shows no geosyncline where the hypothesis requires it in South America. And second, continuity of geosynclinal systems does not require continuity of land. A geosynclinal, mountain-forming system now connects South America (the Andes) and western Antarctica (the Antarctic Peninsula) across an ocean gap. However, this system does suggest a long-standing relationship between the tip of South America and western Antarctica which should be a guide to reconstructions of continental relationships in the past.

Wilson (1963a, b) suggests that secondary (not mid-oceanic) ridges on the ocean bottoms may connect former contact points of continents, and he offers a mid-Mesozoic "supercontinent" assembled according to these guides. The results disqualify the method, I think, so far as the southern continents are concerned. The continents are fitted together in a very unnatural way, with the Antarctic Peninsula rotated far out of its probable relation to South America (see preceding paragraph) and with the southern quarter of South America overlapping the ancient (Precambrian) shield of Antarctica, although this part of South America has its own geologic record running far back into the Paleozoic (Harrington). I do not think Wilson mentions this difficulty.

Geologic structures probably do exist that indicate relative positions of continents in the past. The link between South America and West Antarctica is probably one. However, I know no others like it in the Southern Hemisphere. For the reasons given, I specifically doubt the significance of du Toit's Samfrau and of Wilson's suboceanic guide lines.

Paleoclimatic evidence. Because climate is correlated with latitude (in a general way but very far from precisely), climatic indicators may show something about the positions of continents in the past. Among current publications on the subject is *Descriptive Palaeoclimatology* (Nairn 1961). This is a stimulating book, but before being too much influenced by it readers should consult also Bucher's (1962) review of it. Other important recent books include Schwarzbach's (1963) *Climates of the Past*, Ager's (1963) *Principles of Paleoecology*, Imbrie and Newell's (1964) *Approaches to Paleoecology*, and Nairn's (1964) *Problems in Palaeoclimatology*.

The distribution of glaciation in the past, the nature and distribution of fossil floras, and the distribution of some fossil faunas, especially of shallow-water marine animals, are valid criteria for mapping ancient climates. Other kinds of evidence, including the nature of ancient soils and the distribution of deposits of salt and other "evaporites," are more difficult to interpret, and interpretation often suffers from what Bucher calls "perplexing ambiguities" and the "palaeomagneticist's overvaluation of latitude as a climatic factor," as well as from insufficient data. I doubt if the nonglacial, nonbiogeographic evidences of ancient climates yet provide good arguments for or against continental displacement, and I shall discuss them no further here. I think this is true also of diversity gradients in the distribution of fossil floras and faunas (Stehli and Helsley 1963; Irving and Brown 1964). These gradients probably reflect gradients of climate in some cases, but they are too complex and too little understood to be worth discussing here.

Permo-Carboniferous glaciation. Ice sheets on land gouge rock surfaces, deposit thick glacial "drift," and move boulders far from their places of origin. Evidences of major glaciation may persist for hundreds of millions of years, and all together they are unmistakable, although turbidity currents, redeposition of glacial debris, and some other nonglacial agents may counterfeit some effects of glaciation.

Evidences of continental glaciation occur in the late Carboniferous and early Permian, and the distribution of ice then is remarkable (Fig. 30). Ice sheets were apparently widespread in southern Africa north into what is now the edge of the tropics, and in southern Madagascar; in southern South America north into the present edge of the tropics, although the ice may not have reached the extreme southern tip of the continent (but the Falkland Islands were glaciated); and in the southern half or more of Australia from the extreme south including Tasmania north into the present edge of the tropics (but New Zealand was apparently not glaciated at this time—see p. 104). Signs of Permo-Carboniferous glaciation have been found on Antarctica too, al-

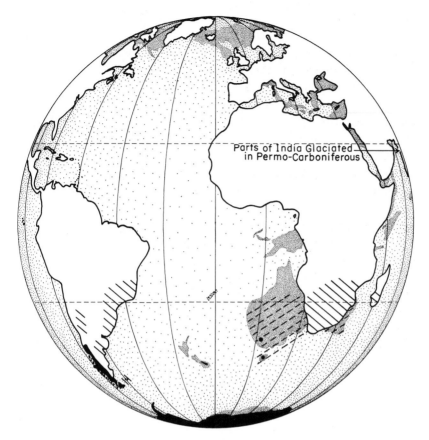

Fig. 30. Distribution of glaciation in the Southern Hemisphere in the Permo-Carboniferous (*hatched areas plus Antarctica*) and in the Pleistocene (*solid black*). Diagrammatic: ice sheets were not necessarily continuous within the areas indicated.

though the extent of it has not been determined (because the present ice cap covers the marks of the earlier ice). And India, which now lies mainly in the tropics *north* of the equator, was apparently heavily glaciated. But more-northern continental areas were apparently not glaciated then.

The wide distribution of late Carboniferous or early Permian ice has in general been confirmed by local geologic studies on each southern continent (Chapters 8–12), although the centers and directions of movement of the ice are still doubtful in some cases. The glaciations of the different areas were approximately contemporaneous, but the times cannot be correlated exactly, and some persons think that different pieces of land were glaciated

successively. I cannot evaluate either the evidence of succession
or its significance, if succession occurred. The term "Permo-
Carboniferous glaciation" includes all the ice sheets listed above
regardless of exact chronology. See Kummel (1961:328–347) for a
longer but still concise summary of this glaciation and related
"Gondwana formations."

The Wegenerian explanation of the distribution of glaciation
in the Permo-Carboniferous is that the southern continents and
India then lay somewhere near the South Pole. However, this
explanation is not accepted by all climatologists. Coleman (1926),
after a comparison of recent and ancient ice ages, thought the
distribution of Permo-Carboniferous glaciation could be explained
without moving the continents. He did not decide what the explanation actually was, but suggested a somewhat indefinite "conjunction" of astronomic, geologic, and atmospheric conditions.
Brooks (1949:248, Fig. 29; my Fig. 31) drew an actual plan of
land and water that he thought might have resulted in Permo-
Carboniferous glaciation without changing the positions of continents. However, his plan requires extraordinary, improbable
changes in the shapes of continents and in intercontinental con-

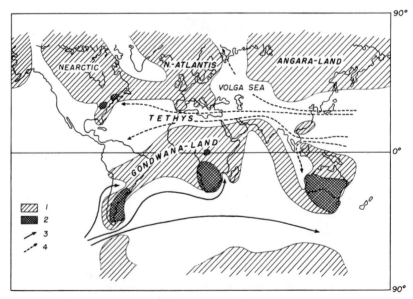

FIG. 31. Brooks's hypothetical "geography of the Upper Carboniferous" (from
Brooks 1949, Fig. 29, redrawn and simplified): 1, land; 2, ice sheets; 3, cold
currents; 4, warm currents.

nections, and the results of the changes are doubtful. As Schwarzbach (1963:150) says, it is hard to see why Brooks's Gondwanaland, in a position near the equator, should have been the site
of ice sheets. Brooks's plan is contradicted by new paleomagnetic
evidence, as I have noted at the end of Chapter 14. And the
plan fails to account for the glaciation of India. I think (speaking
again as a layman) that Coleman's indefinite explanation and
Brooks's more detailed plan are equally unsatisfactory, and I see
no reasonable explanation of the distribution of the old glaciation except that the continents lay far southward then. On the
other hand, I doubt if the glaciation would have occurred if the
southern continents were actually united. The interior of an enormous supercontinent centered near the South Pole would probably receive too little precipitation to form ice sheets and to support the flora that accompanied and followed the ice. Glaciation
such as did occur would be most likely if the southern continents
and India lay near the pole but were separated by areas of ocean
sufficient for massive evaporation. The arrangement of land suggested in Fig. 38, which will be the basis of discussion in much
of the rest of this book, seems consistent with the distribution of
glaciation in the Permo-Carboniferous. All principal areas glaciated then are brought near or within 45° of the South Pole, into
latitudes where glaciation is possible, judging by what happened
in the Northern Hemisphere in the Pleistocene. The suggested
arrangement includes ocean gaps from which evaporation can
occur to supply water for the ice on land. And the eccentric
position of Antarctica seems greatly to increase chances of glaciation on surrounding continents. A massive ocean current would
probably circle Antarctica from west to east. Such a current is
probably always present in the Southern Hemisphere when the
land allows it. With the land arranged as suggested, the current
would be cooled by passage through the polar region and would
then be deflected away from the pole toward the continents that
were in fact glaciated.

The direction of movement of ancient ice sheets is more difficult to determine than their distribution, for evidences of direction are sometimes ambiguous. However, some persons have tried
to show that Permo-Carboniferous ice moved outward in a radial
pattern from a main center on Antarctica onto other continents
that were supposedly then in contact, including Australia and

Africa. This pattern of movement is diagrammed by King (1961:
310, Fig. 1; my Fig. 26) and would be almost conclusive evidence
of continental contact, if the diagram were based on fact. I am
not a glaciologist and I cannot discuss this matter authoritatively.
However, I have tried to ascertain the facts especially in the one
pertinent place that I know best, which is Tasmania.

King's diagram indicates movement of Permo-Carboniferous
ice northward across Tasmania into Australia. Such movement
has been accepted as fact by earlier Australian geologists, including David (1950:I, 315), who thought that movement of Upper
Carboniferous ice was northward across Tasmania and that ice
"must have come from high land lying to the south of Tasmania."
But in the more recent *Geology of Tasmania* (Spry and Banks 1962)
the evidence is re-examined and the conclusion (p. 214) is star-
tlingly different:

> Late in the Carboniferous Period or early in the Permian ice covered
> much of Tasmania . . . The glaciers from an ice center north-west of
> Zeehan diverged about a higher area near Cradle Mountain. One tongue
> occupied a deep valley near Wynyard and a lobe fanned out south of
> the high area to occupy parts of northern and central Tasmania.

Erratics from western Tasmania were dropped in eastern Tasmania. Parts of Tasmania may even have been ice free. Glaciation of Tasmania thus seems to have been from a local center,
and the direction of ice movement was from west to east and
south, *not* from south to north. Apparently no great ice sheet
moved northward across Tasmania, and this fact seems to me to
be fatal to the idea of radiation of ice from one main antarctic
center.

After reaching this conclusion about Tasmania, I have tried to
ascertain the facts in South Africa too. The direction of movement of Permo-Carboniferous ice in South Africa, shown partly
by distribution of erratics carried by ice from known sites, is discussed and mapped by du Toit (1954:276–277, Fig. 4). He concludes that there were several ice centers, that they were far
north of the southern tip of the continent (near the present southern edge of the tropics), and that the ice moved toward the (present) south and southwest.

Although (so far as I know) no one recently has disputed the general southward movement of most Permo-Carboniferous ice on South Africa, Haughton (1952:255) suggests that

variations in lithology and in thickness of the Ecca beds in the portion of the Karroo basin which lies south of approximately 30° 30′ S. latitude seem to show that the material of which these beds are composed was derived from a land-mass lying to the south, which also provided most of the material for the southern facies of the Dwyka tillite.

Haughton himself later (1963:201) summarizes his conclusions by saying that the most southern glacial material "may well" have been derived from a glaciated area south of the present coast of Africa. King (1962:41), however, referring to Haughton's earlier paper, says that the latter "prescribes" an ice source south of Africa, and this is apparently the entire evidence on which King (1961:310, Fig. 1; 1962:42, Fig. 16) shows ice moving onto the southern tip of Africa from the south. I do not have the technical knowledge to criticize the evidence, but the picture produced in my mind of two great ice sheets meeting head-on in South Africa is difficult to accept. Perhaps an explanation is possible, but I do not think either Haughton or King has offered one.

The directions of ice movement shown by King on India and South America are doubtful and disputed (Schwarzbach 1963: 140, 147–148; my p. 111). And the directions of ice movement given by King for Antarctica are the directions now. Directions of movement of Permo-Carboniferous ice on Antarctica are actually unknown, and (if the continents have drifted) need not have been the same as now.

In short, King's diagram of supposed radiation of Permo-Carboniferous ice seems to be virtually without factual basis.

I have criticized this diagram in detail because of what might be called its reverse significance. The diagram shows the pattern of ice movement that Wegenerians think ought to be found if the southern continents were united during Permo-Carboniferous glaciation. And I agree, up to a point. Enormous ice sheets did occur on southern Africa, southern South America, southern Australia, India, and probably on Antarctica. The great amount of the ice suggests (I think) that the continents were not actually united, but if they were united, a radial pattern of movement something like that diagrammed by King would be expected, and the evidences of it should exist and should be unmistakable. The pattern does not exist, or at least is not unmistakable. This seems to me to be real evidence that the continents were not united when the glaciation occurred.

My conclusion, as a deeply interested but technically unqualified critic, is (1) that the distribution of Permo-Carboniferous ice very strongly suggests but does not quite prove that the southern continents and India lay far southward, but (2) that the directions of ice movement are insufficiently known and that movements of ice on different southern continents probably did *not* form a significant pattern of spreading from a common center. Both the amount of ice on land (suggesting evaporation from intervening water gaps) and the absence of an unmistakable radial pattern of ice movement seem to me to suggest that the southern continents (excepting Africa and South America) were not actually united but lay in a pattern something like that suggested in Fig. 38.

Biogeographic evidence. If there is such a thing as a professional in biogeography, I am one. I am therefore in a position to know the complexities and difficulties of the subject. They are many.

Distributions of terrestrial organisms reflect both distribution of land and distribution of climate, as well as other factors, including distribution of competitors. The effects of different factors on distributions of plants and animals in the past are sometimes difficult to distinguish, and some biogeographers even fail to distinguish the effect of present climate from that of past geography. This error is common among Wegenerians.

What I said some years ago (1949) about Wegenerian zoogeography is worth repeating.

Several usually unstated assumptions underlie this book [Jeannel's *La Genèse des Faunes Terrestres* . . .] and other Wegenerian zoogeography in which the method is to match distributions of existing animals against patterns of hypothetical ancient lands. First, it is assumed that animal distributions are more permanent than land; in the present case, that insects have moved less than continents. The whole method depends on this . . . Another assumption is that ancient geography is the *only* important factor governing animal distribution; that existing climate, for example, can be ignored . . . A case in point from the present book

[Jeannel's, pp. 250–251] is the carabid genus *Morion*. This genus is now pan-tropical. The main existing tropical regions are about equal to ancient, hypothetical Inabresia, so *Morion* is stated to have evolved on Inabresia and to be of Jurassic age, and it is noted as extraordinary that species of *Morion* have differentiated so little since then. That *Morion* may be limited northward and southward by existing cool climates is ignored—if its range is determined by existing climate, how can it be supposed to reflect the shape of ancient land? Still another Wegenerian assumption is that land animals never or rarely get across even narrow water barriers between continents, so that widely distributed animals must (by Wegener's time-table) be very old. But placental mammals, at least some rodents and many bats, have spread over every continent during the Tertiary. In the face of this fact, why is it necessary to go back to the Jurassic to account for the wide distribution of insects such as *Morion?*

Besides the difficulties stated or implied in this quotation, there is often doubt about the climatic requirements and tolerances of ancient plants and animals, about their means of dispersal, and (because the fossil record is incomplete) about what their distributions really were. Nevertheless, the distributions of appropriate plants and animals do give useful information about the world in the past.

Evidence of existing plants and animals. For reasons given above, I doubt if the distribution of plants and animals now has anything to do with continental drift. My point is not that existing distributions are incompatible with drift, but that they are not evidence of it. They are simply noncommittal. Most existing groups of plants and animals, regardless of their times of origin, have or may have dispersed or redispersed since the latest time of continental contacts that is allowed by other evidence. Existing mammals (Chapter 7) and probably also fresh-water fishes (p. 81) and angiosperms (p. 167) are among the organisms that have spread over the world too late to have been dispersed by continental drift. If some existing plants or animals do preserve distributions that date from a time of possible continental contacts, alternative explanations of their distributions are possible. What I shall say of the conifers can be said of many other groups of plants and animals: their distributions are consistent with any of several paleogeographic hypotheses, and so are proof of none.

Distribution and history of conifers. Conifers (pines and their allies) and taxads (yews, and so forth) are an example of a group of plants of which both the present distribution and the fossil record have been cited as evidence of continental drift. Conifers have existed at least since the late Carboniferous; taxads, since the mid-Mesozoic. These plants dominated forest vegetations during much of the Mesozoic, and they are still widely distributed and still dominant in some parts of the world. Their geographic history is reasonably well known and has been reviewed by Florin (1963).

These plants seem to have been distributed in two distinct zones through most of their known history. One set of families and subfamilies, among them the true pines, has occupied the northern part of the world. Another set of families and subfamilies, including the "southern pines" (araucarians, and so forth), has occupied the south, including southern South America, Australia, and New Zealand, and some of the southern groups have been represented also in South Africa and India. The general separation of northern and southern conifer-taxad floras since the Carboniferous is remarkable, although there have been a few exchanges.

Many phytogeographers, including Florin, explain the zonation of conifer-taxad floras as due to former separation of northern and southern continents by physical barriers, including the Tethys Sea, and to former land connections in the south. However, conifers and taxads were not alone in the world, and the distribution of some other, contemporaneous plants and animals shows that northern and southern land areas cannot have been completely separated (Chapter 7). Partial or temporary barriers probably did exist between the northern and southern continents, but many plants and animals crossed them. Why did not conifers and taxads do it? Or rather, why did not more of them do it, for a few did? I think the answer must be that conifer-taxad distribution has depended less on distribution of land than on climate.

The present distribution of these plants fits my second pattern of climatic zonation (Chapter 14) of opposition of the north temperate zone to the rest of the world. Northern conifers and taxads, which live in strongly seasonal climates with periods of intense cold, do not extend far into the tropics. But the southern groups, although they are now best represented in the far south, often

do extend deeply into the tropics or are widely scattered as relicts there. The *present* distributions of *Araucaria* and *Podocarpus*
(Figs. 32 and 33) illustrate this fact, and so do the distributions
of other groups mapped by Florin.

The known distribution of northern and southern conifers and
taxads from late Carboniferous to Eocene times is summarized by
Florin (p. 265) in one map (my Fig. 34). The map gives the impression that these plants were zoned even more strictly in the
past than now, and that the zones were separated by a wide gap
in the tropics, at least in Africa and America. However, this gap
may represent inadequacy of the fossil record in the tropics rather
than an actual gap in distribution of the plants. We do not know
what groups of conifers and taxads were or were not in the tropics
in the past, and our ignorance allows two reasonable supposi-
tions. The first is that zonation of these plants may have been
the same in the past as now, with one set of families or subfamilies occupying the north and another set occupying the tropics
and far-southern areas. And the other supposition, which follows
from the first, is that the southern groups may have dispersed
through the tropics rather than by far-southern routes, and may
then have died back into southern relict areas, perhaps as a result
of competition with evolving angiosperms. The restricted, discontinuous, relict distributions of many southern conifers and
taxads are consistent with such a history.

What do conifers and taxads really tell about the state of the
world in the past? I think they do not prove existence of old land
connections either in the far south or elsewhere. Florin confirms
occurrence on the Antarctic Peninsula (or Archipelago) at least
of araucarians in the Jurassic and early Tertiary and of a podocarp (*Acmopyle*) in the early Tertiary, and Barghoorn (1961) records fossil leaf impressions of conifers probably "of araucarian
type" on Antarctica about 3° from the South Pole. However,
these plants may not have required continuous land to reach
Antarctica, and, if they did, a single link, with South America
would probably be enough. Araucarians and *Acmopyle*-like podocarps are fossil in southern South America at appropriate times
(Florin, maps on pp. 178, 185). If southern conifers and taxads
dispersed through the tropics (see preceding paragraph) and if
Africa and South America were in contact then, the plants presumably spread across the land. However, tropical organisms

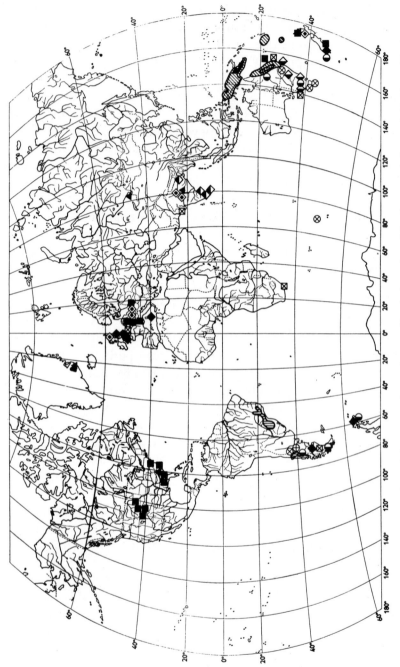

Fig. 32. Distribution of conifers of the genus *Araucaria* now (*hatched areas*) and in the past (*other symbols*) (from Florin 1963, Fig. 14; copyright *Acta Horti Bergiani*).

Fig. 33. Distribution of conifers of the genus *Podocarpus* now (*hatched areas*) and in the past (*other symbols*) (from Florin 1963, Fig. 25; copyright *Acta Horti Bergiani*).

189

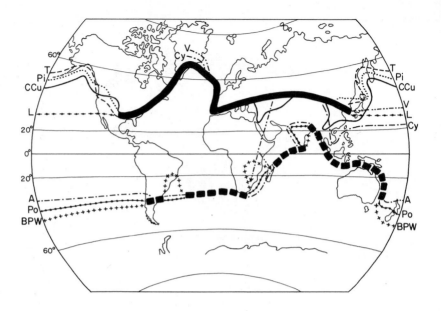

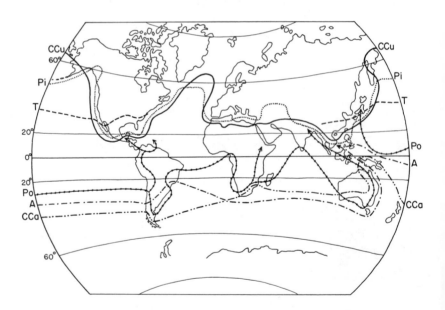

Fig. 34. Distributions of conifers and taxads in the past (*above*) and now (*below*) (from Florin 1963, Figs. 64, 65, redrawn and simplified). The past distributions cover late Carboniferous to Eocene time. Lines show approximate southern limits of northern groups and northern limits of southern groups. These maps are diagrammatic, to emphasize major patterns. See Florin for further details and for explanation of symbols.

have been able to disperse around the world by other routes more recently. The distribution of early conifers is not good evidence of continental contracts.

Conifers and taxads probably tell more about ancient climates than about land connections. Florin (pp. 265, 275) summarizes and compares the past and present distributions of these plants in two maps (my Fig. 34). The maps, though different in detail, are very similar in general pattern and suggest that the climatic zones of the world have changed very little since the late Carboniferous. This is hardly consistent with current theories of continental drift. However, the map of older distributions is probably (necessarily and excusably) based on incomplete data and oversimplified. It should be interpreted with caution.

The distribution and history of conifers and taxads, as now known, can probably be reconciled with any reasonable history of the world: stability of continents, or continental drift, or antarctic land bridges. The distribution of these plants seems consistent with any of these hypotheses, and proof of none.

Permo-Carboniferous floras. Two very different, major floras existed in different parts of the world during the late Carboniferous and Permian periods. Both floras formed coal and have left extensive fossil records. Their distributions are diagrammed in Fig. 35; they are in general well known and have been discussed many times (for example by Ager 1963, Schwarzbach 1963, and Kräusel 1961), although some details are still doubtful. One flora was chiefly southern and was very widely distributed in the far south. It accompanied or followed Permo-Carboniferous glaciation wherever the glaciation occurred—on South Africa and Madagascar, southern South America, southern Australia and Tasmania, and Antarctica—and this flora occurred on India too. It was a relatively species-poor flora, dominated by *Glossopteris* and allied genera. Fossil tree trunks found with it in India (Seward 1931: 244–245, Fig. 73) as well as on the southern continents show prominent annual growth rings, indicating a strongly seasonal climate. Such rings are usually formed now as a result of alternation of summers and winters in cold latitudes. Some of the species in the *Glossopteris* flora were very widely distributed, being apparently identical on different pieces of land (listed above) that are now widely separated. For example, more than half of the

Fig. 35. Distribution of major floras in the late Paleozoic (from Ager 1963, Fig. 17.3; copyright McGraw-Hill Book Co., Inc.).

species on Antarctica apparently occurred also on India (p. 111). These plants are of course known only as fossils, mostly leaves, so that absolute identities are difficult to prove, but the apparent identity of so many species on continents that are now widely separated is impressive.

At the same time that the *Glossopteris* flora existed on the southern continents and India, a different flora, more diverse (consisting of more species of plants), and with different adaptations existed in parts of what is now the north temperate zone. This was the flora that produced the great coal beds of the north. Its climatic requirements are not exactly known, but it apparently was a swamp-forest association, consisting of plants structurally adapted to a relatively warm, uniform, humid climate.

This geographic differentiation of floras in the late Carboniferous and Permian surely reflects a climatic difference between the southern and northern parts of the world. The significance of it will be further considered in the following chapters.

Wegenerians continually assert, quoting each other to prove it, that *Glossopteris* and its allies could not cross ocean gaps but must have dispersed over continuous land, and that all the land areas where the plants occurred must have been united. This was a poor argument to begin with, considering how many other plants have reached islands across oceans, and little is left of the argument now that the fruiting structures of *Glossopteris* are known. The fruiting structures of at least some *Glossopteris* and allied forms (Plumstead 1956; my Fig. 36) were leaflike and have been compared to sail planes (Plumstead, p. 21), and the seeds themselves were apparently very small, about 0.5 mm in diameter in some cases. The plants may have been dispersed partly by wind, and, since they were frequently associated with glaciation, they may have been carried by floating ice too. I do not pretend to know how they really did disperse, but their distribution is not good evidence of continuity of land. On the other hand, terrestrial plants and animals are much more likely to get across narrow water gaps than wide ones, and the distribution of the *Glossopteris* flora does suggest that gaps between southern continents were narrower then than now. I think dispersal of this flora might have occurred across the water gaps indicated in Fig. 38, especially if strong winds and ocean currents circled Antarctica then as they do now.

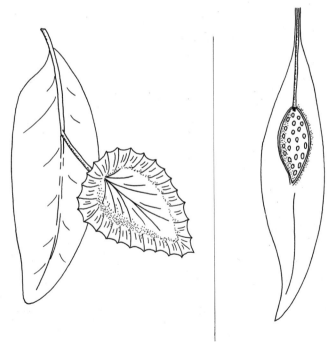

Fig. 36. Fruiting structures of *Glossopteris* (from Plumstead 1956, Figs. 1b and 4b, redrawn).

Permian and Mesozoic reptiles. Zoogeographically, terrestrial reptiles were the most significant land animals before the Tertiary. Reptile faunas of northern and southern continents were apparently closely related in the late Permian and early Mesozoic (p. 83, Fig. 16). Some Mesozoic groups of reptiles, including dinosaurs, occurred on all the continents that are habitable now, indicating that these continents were all somehow accessible to land life then. But there is no evidence that any terrestrial reptiles reached Antarctica or that any Mesozoic ones except *Sphenodon* reached New Zealand. *Sphenodon* probably crossed a water gap to reach New Zealand (Chapter 10). Finding of similar animals fossil on Antarctica (if they should be found—they have not been) would not be proof of land connections.

Dinosaurs. Dinosaurs are known to have occurred on every continent except Antarctica through much of the Mesozoic, and on Madagascar, but not on New Zealand. However, different groups of them had different distributions, and during the later Cretaceous the distribution of plant-eating groups conformed to the present climatic zones. Sauropods were most numerous in the tropics (although some of them occurred in northern regions too), while ornithopods and ceratopsians occurred only or chiefly above the tropics in Europe, Asia, and North America (see Colbert 1961:272–274, for geographic lists of Upper Cretaceous dinosaur genera). This distribution indicates that a north temperate climatic zone was differentiated then and that some dinosaurs not only tolerated less-than-tropical conditions but were specially adapted to them.

Bridge-building zoogeographers have made much of the wide distribution of certain dinosaurs in the mid-Mesozoic. The animals in question were gigantic sauropods. Their remains have been found in Africa, Madagascar, India, South America, and Australia (as well as farther north in Eurasia and North America), and this distribution has been thought to require complete land connections, preferably direct ones between the southern continents. However, these dinosaurs probably did not live on land but mainly in water. They seem to have been adapted to breathe while submerged, with only their nostrils exposed at the water surface (Colbert, pp. 102–104). They may have been better swimmers than hippopotamuses. A hippopotamus (not the common existing African species) is fossil in the Pleistocene of Madagascar and evidently reached the island across an ocean barrier that few other mammals have crossed, and sauropod dinosaurs may have crossed still wider barriers. Animals that live in fresh water, for example fresh-water fishes, usually spread over the world more slowly than terrestrial animals do, unless the fresh-water forms have special dispersal mechanisms or can go through salt water. I think, therefore, that the wide distribution of amphibious sauropod dinosaurs is not evidence of land connections but probably means that the animals could enter salt water and that they dispersed along coast lines and across at least narrow ocean gaps.

Summary of reptile evidence. Some details of the distribution of reptiles in the past are at least consistent with the hypothesis of continental drift. For example, the close relationship of South African and South American reptile faunas in the mid-Triassic is consistent with actual union of the continents, as is the distribution of mesosaurs early in the Permian (p. 169). The occurrence of dinosaurs on Madagascar would be consistent with land connections there too. But the real significance of these details is doubtful. The record of reptiles as a whole seems to me to suggest a general accessibility of all now habitable continents and isolation of Antarctica and New Zealand rather than specific southern land connections.

Shallow-water marine faunas. The distribution and significance of shallow-water marine invertebrates, now and in the past, are complex and difficult subjects which I am not competent to treat in detail. In general, the distribution of animals in the ocean, even in shallow water, probably reflects distribution of temperature more than distribution of land, but the temperatures that ancient forms required are doubtful. For example, fossil corals have often been cited as proof of tropical climates in the past, but two different groups of corals are now distinguished, one of which can tolerate cold water, so that the significance of some fossil coral reefs must be reappraised (Craig 1961:215).

The significance of the distribution of fossil marine invertebrates depends partly on the widths of deep water the animals are likely to have crossed, which depends on the lengths of their larval periods. Too little is known about this. Thorson (1961:455–474) discusses the complexity of this matter, the various ways in which larval periods can be or may be lengthened, and the occurrence of "long-distance larvae" in some cases. The diversity of larval periods now should be a warning against setting limits to them in the past.

According to Doumani and Long (1962:180), the fauna of shallow antarctic seas was closely related to that of other regions until the mid-Mesozoic but the similarity has decreased since then, the antarctic marine fauna becoming progressively more differentiated from other faunas. If this trend is clear enough to be significant (which I am not competent to judge), it might be a result of a gradual drifting apart of the continents. Or it might reflect differentiation of climates. Or it might be partly a result of a general evolutionary trend toward shorter larval periods among marine animals. Such a trend would decrease the chance of dispersal of shallow-water forms across deep water and increase endemism of shallow-water faunas. A trend like this would be comparable to complex parallel evolution by "grades" (Mayr 1963:608–609) and might reflect a general advantage in the shortening of larval periods. The advantage of shorter larval periods might itself increase with time, perhaps with the evolution of increasingly numerous or increasingly effective predators feeding on larvae in the sea.

Cloud (1961:151–200), at the end of a paper on "paleobiogeography of the marine realm," concludes that the fossil record of shallow-water invertebrates favors no post-Triassic and probably no post-Carboniferous continental drift. I do not cite this opinion in order to accept it. I should like to base my conclusions on evidence rather than on previous opinions, and I cannot assess this particular evidence satisfactorily. However, Cloud's opinion may be a useful antidote to the assertions of some Wegenerians that the distribution of fossil marine invertebrates does indicate continental drift.

Biogeographic conclusions. Three generalizations can, I think, be made about the significance of plant and animal distribution in the late Paleozoic and Mesozoic.

First, the wide distribution of significant terrestrial groups of plants and animals at various times in the past indicates that all the now-habitable continents were somehow accessible to terrestrial organisms, at least from time to time, from the Permo-Carboniferous to the present, in spite of the existence of the (fluctuating?)

Tethys Sea and in spite of the (temporary?) isolation of South America and Australia. Northern and southern land hemispheres were not so isolated from each other as to stop the exchange of terrestrial life completely for any long period. However, nothing yet found on Antarctica or New Zealand requires land connections at any time, I think. Even the discovery of fossil terrestrial vertebrates on Antarctica (they have not been discovered) would not necessarily be proof of land connections, for certain kinds of terrestrial vertebrates have often reached islands across moderate water gaps.

Second, the earlier distribution of (fossil) plants and animals show patterns of climate, primarily zonations of climate, rather than specific patterns of land. The general accessibility of most continents, summarized above, suggests that land connections existed in some places at some times, but plant and animal distributions as now known do not show where the earlier connections were. Only in the late Cretaceous and especially in the Tertiary do plant and animal distributions begin to show specific land connections and specific ocean barriers (Chapter 7), and this is too late to be significant in any likely hypothesis of continental drift.

Third, in spite of the preceding generalizations, one specific piece of biogeographic evidence does suggest changes in the positions of continents. This is the distribution of the *Glossopteris* flora in the Southern Hemisphere during and after Permo-Carboniferous glaciation. The geographic limits of this flora were probably determined by climate rather than by physical barriers, and the plants concerned could probably cross considerable water gaps. Nevertheless the distribution of this flora does strongly suggest that most continents lay farther south then than they do now, and the apparent identity of many species on continents that are now very widely separated strongly suggests that southern water gaps were narrower then than they are now.

20. Permo-Carboniferous and Mesozoic history (3): paleomagnetism; comparison of evidence

Paleomagnetism. The most impressive new evidence that continents may have moved is derived from remanent magnetization (paleomagnetism or "fossil magnetism") of rock. Some rocks, including both some igneous and some sedimentary formations, contain ferromagnetic constituents that were aligned by the earth's magnetic field when the rocks were formed. These rocks have a fixed magnetization of their own which points where the poles were at the time of origin of the rocks—*if* certain assumptions are correct.

The paleomagnetic pointers may be compared to compass needles frozen into ice, which show the direction of the (magnetic) pole when the ice was formed no matter how the ice may have drifted and turned afterward. The frozen needles may show also, by their dip, how far away the pole was. And the direction plus the distance give the position of the pole when the ice was formed, or rather show the initial relation of the ice to the pole. If the pole has been fixed and the ice has changed in orientation or distance from the pole, the change will be shown by divergence of the frozen compass needles from the pole now. Or if the ice has been fixed and the pole has moved or "wandered," the needles will show this change in the same way. But the information the frozen needles can give is limited. They cannot show whether it is the ice or the pole or both that have moved but only that a change has occurred in relation of ice to pole. And they cannot show movements of the ice in a circle around the pole, if the distance from the pole and the orientation of the ice to the pole have not changed.

This analogy is an oversimplification. The rocks used in paleomagnetic work do not contain visibly aligned particles. They are directionally magnetized, but their magnetism is usually weak and often complex, the direction of it must be detected by special instruments, and the poles must then be found by mathematical formulas. Nevertheless the analogy is exact as far as it goes. The invisible magnetic pointers "frozen" into rocks when the rocks are formed may show as much about the history of a continent as frozen compass needles might show about the history of an iceberg. And the limits are the same. The limits of paleomagnetic evidence are important and are further discussed in the fifth and sixth following paragraphs.

The basic theory of paleomagnetism is complex and difficult. For a clear, not too technical treatment of it, see the short book on magnetism by Parasnis (1961:106–121). For a more detailed treatment, see Cox and Doell (1960). And see Irving (1964) for discussion of the applications as well as the theory of paleomagnetism. Both Cox and Doell (1960) and Irving (1964) give long tables of actual paleomagnetic determinations. These tables represent the common pool of data that all paleomagneticists use.

Possible sources of error are so numerous that determinations from single samples are not dependable, but when samples from different parts of one rock formation and from different contemporaneous formations on one land mass are combined, errors average out and the results become significant. This is proved by results from relatively recent (Quaternary) deposits, which place the paleomagnetic pole close to the present north pole of rotation (Parasnis, p. 113, Fig. 31).

In the more distant past, the paleomagnetic pointers seem to show the pole very far from where it is now, and the pointers on different continents seem to show it in different places. Actual findings can be coordinated in various ways, for example as maps of "virtual" pole positions or diagrams of apparent polar wandering, different for each continent (Parasnis, p. 118, Fig. 33), but the method of coordination that is pertinent now is to move the continents until the paleomagnetic pointers of all of them, in rocks of a given age, point to the existing north (or south) pole. This requires moving some continents long distances from their present positions, and suggests that the continents have moved from the old positions. Movements of Australia suggested by paleomagnetic data are diagrammed in Fig. 37. An attempted reconstruction of land in the Southern Hemisphere in the mid-Mesozoic, suggested partly by paleomagnetism, is shown in Fig. 29.

Four questions should be asked about paleomagnetism. First, is it theoretically above suspicion as evidence of continental movement? It is not. Paleomagnetic reconstructions depend on the

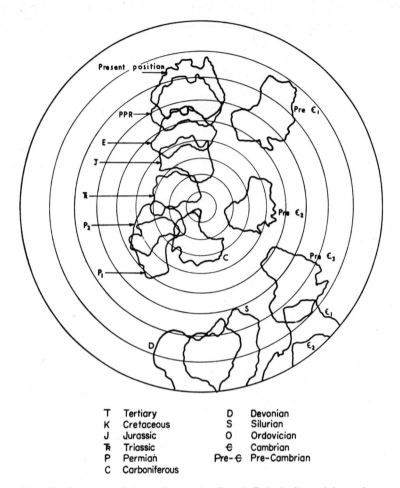

T	Tertiary	D	Devonian
K	Cretaceous	S	Silurian
J	Jurassic	O	Ordovician
Ŧ	Triassic	Є	Cambrian
P	Permian	Pre-Є	Pre-Cambrian
C	Carboniferous		

Fig. 37. Relation of Australia to the South Pole indicated by paleo-magnetism (from Runcorn 1962, Fig. 22; copyright Academic Press). The concentric circles are spaced 10° apart. The paleomagnetic data on which this figure is based do not show longitudes (Runcorn, p.25) and do not require moving Australia around the pole. The relation of Australia to the pole actually changes very little from the Carbonifer-ous to the Jurassic. During this whole time the Tasmanian corner of Australia is placed on or within the Antarctic Circle, which is not shown but would lie outside the second circle from the center of the diagram.

assumption that the earth's magnetic field has always been essenti-ally the same as now, with the magnetic poles at or near the poles of rotation. This is probable but not certain. The earth's magnet-ism is not fully understood and the basic pattern of it may have changed from time to time. For discussion of this question see Cox and Doell (pp. 661ff), Parasnis, and Runcorn (1962, especially pp. 9–10).

Second, what are the theoretical limits of paleomagnetic evidence? There are limits, suggested in the second and third paragraphs of the present chapter. If the basic assumptions are correct, as I think they probably are (because paleomagnetic findings seem to me to agree reasonably well with other evidence), paleomagnetic determinations can show the orientation of a continent and its distance from a pole in the past, and this information is useful and in some cases dramatic. But its usefulness is limited in two ways. Series of paleomagnetic determinations may show changes in the relation of a continent to a pole but do not show whether it is the continent, the pole, or both that have moved. This difficulty may be partly overcome by comparison of the paleomagnetic records of different continents, which may show whether the continents have moved relative to each other or whether the poles have moved in definite paths that are the same in relation to all continents. Actual paleomagnetic determinations do indicate that different continents have moved long distances independently but do not yet clearly show that the poles have moved (Deutsch 1963). "Polar wandering" is therefore still a less probable phenomenon than continental drift (see Postscript).

The other limit inherent in paleomagnetic evidence is that a continent carrying a paleomagnetic "pointer" may have been anywhere around the earth at the distance from the pole and with the orientation indicated. Therefore, although paleomagnetic determinations may show latitudes and orientations of continents in the past, they do not show longitudes. This is what Runcorn (p. 25) means by saying, "Unfortunately the indeterminacy of the ancient longitudes presents a difficulty." The difficulty is discussed and illustrated by Irving (1964:266, Fig. 10.14). This difficulty is not always stressed as much as it should be by Wegenerians attempting to use paleomagnetic data to reconstruct ancient supercontinents. The fact is that *paleomagnetic determinations do not show longitudes,* or at least do not show them directly. This difficulty too may perhaps be overcome eventually by comparison of the polar-wandering curves of different continents. This hypothetical method of determining relative longitudes in the past requires that the poles have in fact wandered, which is still doubtful (see above), and requires also many more actual determinations of sequences of pole positions than are yet available for most continents. Recent attempts by Creer (1964a; 1964b) to reconstruct Upper Paleozoic and mid-Mesozoic positions of southern continents by this method seem intended to illustrate what may be possible in the future rather than what can be done now (Creer 1964a, p. 1120, last paragraph). (See my Postscript.)

These limits to the use of paleomagnetic evidence are important and should be remembered. However, they should not be over-emphasized. In spite of the limits, the new science of paleomagnetism is giving us an exciting new view of the past. The paleomagnetic record is counteracting old prejudices against the idea of continental drift and, in conjunction with other evidence, may eventually enable us to reconstruct the history of the continents in detail and with confidence.

The third question is, what specific evidence, derived from paleomagnetism, is actually available? For relatively recent, post-Eocene time, paleomagnetic findings are numerous and consistent and seem to show little change in positions of continents (Cox and Doell, pp. 736ff, 756ff). Stability of the continents during this time is, of course, consistent with much biogeographic evidence, which need not be given here. For Mesozoic and early Tertiary (Eocene) time, paleomagnetic results are fewer, scattered, and inconsistent. Large amounts of drift are suggested for some continents (see again Runcorn's tentative reconstruction of the Southern Hemisphere for mid-Mesozoic time, my Fig. 29), but the evidence is scanty and in part doubtful (Cox and Doell, p. 762). For the Upper Carboniferous and Permian, relatively satisfactory pole positions have been determined for three continents: Europe, North America and Australia (Cox and Doell, pp. 760ff). Only "very preliminary" determinations are available for Africa and South America (Irving, p. 225) and none at all for Antarctica and India during the time in question, which was the critical time of late Paleozoic glaciation. The Australian determinations are technically excellent and consistent as well as dramatic, and place the Tasmanian corner of Australia near the South Pole in both the Carboniferous and the Permian, and also in the Triassic and Jurassic (see Cox and Doell, Figs. 24, 25, 28, 30; Runcorn, Fig. 26; my Fig. 37). This paleomagnetic placing of Australia during and after the late Paleozoic gives the best available basis for comparing the evidence of paleomagnetism with evidence from other sources. Lack of satisfactory paleomagnetic "fixes" for Antarctica and India at the time of Permo-Carboniferous glaciation is unfortunate. In India, no rocks have yet been found suitable for paleomagnetic determinations between the Cambrian and the Jurassic (Athavale, Radhakrishnumurty, and Sahasrabudhe 1963)!

The fourth question is, of course, does paleomagnetic evidence agree with other evidence of the positions of continents in the past? In trying to answer this question I find myself in a dilemma. I do not want just to cite the conclusions of other persons. I want to interpret the evidence for myself. But interpretation of paleomagnetic evidence is difficult—authorities disagree basically about it—and I have no special competence for it. I shall, however, do the best I can to fit paleomagnetism into a general comparison of evidence from all sources. In doing this, I shall probably alternate between overtimidity and overboldness and be criticised for both.

Comparison of evidence from different sources. Two general difficulties complicate the task of coordinating evidence and reaching conclusions about the history of the continents. First, the history may have been very complex, not reducible to any simple model or simple explanation. And second, the evidence itself is difficult to assemble, assess, and compare, and this difficulty is increased by continual interposition of the human factor. Most writers try to present detailed factual evidence fairly, but then their arguments and conclusions are sometimes exaggerated, and conclusions offered as opinions by one author may be quoted by later authors as if the opinions were facts. [A cynic, if asked, "How can I prove anything about continental drift?", might answer "Get someone to say it is so, then quote him."] I shall try to make reasonable allowance for these human tendencies, in myself as well as in others.

Real evidence for or against continental drift now comes, I think, from only three sources that are really independent of each other. First is the matching or failure to match of shapes of continents, with supplementary geologic evidence: correspondence or noncorrespondence of strata and other structures on the edges of continents that may have been in contact, the structure of ocean bottoms between continents, and related phenomena. Second is the distribution and pattern of movement of Permo-Carboniferous glaciation, with the distribution of ancient floras. Both the glaciation and the floras as now known show (I think) distribution of climate in the past rather than specific patterns of land. And the third independent source of evidence for or against continental drift is, of course, the record of paleomagnetism.

Africa and South America: paleomagnetic inferences. The relation of South America to Africa is fundamental in the theory of continental drift, and is discussed in Chapter 18. The matching of the edges of these continents and their relation to the mid-Atlantic Ridge strongly suggest, but do not quite prove, that the two continents were once united and have drifted apart. Unfortunately, this case, in which other evidence of lateral (east-west) displacement is relatively strong, apparently cannot be confirmed directly by paleomagnetism. However, Creer (1958) has been able to apply paleomagnetic data to it, even though, as I have said above and as Creer says (p. 388), the data do not show east-west positions. If Africa and South America were once joined together, they have not only drifted apart but have rotated, so that once-parallel coast lines now diverge southward at an angle of about 45° or 50°. Creer's paleomagnetic data seem to show that the divergence was already about 22° in the early Jurassic, and he concludes:

> It seems likely that the movement of the southern continents [since the break-up of du Toit's version of Gondwanaland] into their present positions was more than half completed by the early Jurassic.

However, the break-up of du Toit's hypothetical supercontinent involved northward movements as well as rotations of Africa and South America. Creer's early Jurassic data do not show the supposed northward movements half completed, but place Africa and South America at approximately their present latitudes. Moreover, if separate rotations of Africa and South America were half completed early in the Jurassic, actual parting of the continents (if they were joined) presumably occurred earlier, much earlier than some Wegenerians think (see again Chapter 18). In a later paper Creer (1964a) compares the polar-wandering curves (derived from paleomagnetic data) of Africa and South America and concludes very tentatively that these continents probably separated just before the end of the Paleozoic, in the Upper Permian (but see my Postscript).

Although paleomagnetism does not show east-west positions or displacements of Africa and South America directly, a connection between the earth's magnetic field and the history of these continents is at least possible. The main part of the earth's magnetism is thought to be generated by convection in the core. And the most likely mechanism of continental displacement (if displacement has occurred) is convection in the mantle. Convection is due in both cases to uneven distribution of heat, and release of heat by core convection may determine the pattern of convection in the mantle (see Runcorn, p. 30). If so, deductions about each convection system can be made from the other. The fact that the earth's magnetic field is now virtually parallel to the axis of rotation probably means that core convection is similarly oriented, and if convection in the mantle is correlated with it, east-west movements such as seem to have separated Africa and South America would be expected. On the other hand, if Africa and South America were in contact until (say) the end of the Paleozoic and separated then, the pattern of convection in the mantle probably changed then, and this might be correlated with a change in core convection. And this in turn might be correlated with Cox and Doell's general conclusion that paleomagnetic results are relatively consistent in the Carboniferous and Permian, scattered in the Mesozoic and Eocene, and consistent again but in a different pattern since the Eocene.

Paleomagnetic evidence of union of southern continents. Evidence from other sources (Chapter 19) suggests that most continents lay farther south on the world late in the Paleozoic than they do now but does not show that the southern continents were united. Paleomagnetic evidence too suggests more-southern positions for some continents in the past but does not show actual unions. Because paleomagnetism does not show longitudes, it does not fix the exact positions of continents and would not be expected to demonstrate actual unions in most situations, but it might do so in special cases. If, when better known, the paleomagnetic record should indicate that several continents once lay so far south that they had to be packed together at the South Pole, the existence of a united supercontinent would be strongly suggested. But the record is still much too incompletely known to suggest anything like this during the Permo-Carboniferous.

Creer (1964a) has tried to use comparison of paleomagnetic polar wandering curves to determine positions of continents in the past. The title of his paper is "A Reconstruction of the Continents for the Upper Palaeozoic from Palaeomagnetic Data," but the promise of the title is hardly fulfilled. The reconstruction of the southern continents (Creer, Fig. 7; my Fig. 27) is not really based on paleomagnetism but is du Toit's old plan with selected paleomagnetic determinations fitted to it. The polar-wandering curves show, Creer thinks, that "Gondwanaland" did not begin to disintegrate until the very end of the Paleozoic (Upper Permian) and that Africa and South America as well as the other southern continents separated then, and he thinks that the onset and ending of Permo-Carboniferous glaciation were due to rapid movement of the pole across "Gondwanaland." I do not know what degree of confidence can be placed in either the method or the data on which these tentative conclusions are based, but I think the conclusions themselves are probably wrong. They are hardly consistent with the paleomagnetic record of Australia, which seems to show very little change in the relation of this continent to the pole from the Carboniferous to the Jurassic (my p. 133; Fig. 37). Creer's paper is, incidentally, worth reading as an example of the difficulties, complexities, and uncertainties of paleomagnetic work.

Permo-Carboniferous glaciation and paleomagnetic evidence. Satisfactory paleomagnetic determinations in the Southern Hemisphere for the Permian and Carboniferous have been made only for Australia (preceding pages). However, besides confirming the position of Australia far southward at the time of glaciation, the Australian findings allow a more general testing of glacial patterns.

The approximate glacial pole (the geographic center of Permo-Carboniferous glaciation) is indicated by G on several of the hypothetical supercontinents suggested by Wegenerians (my Figs. 24 to 28), and the paleomagnetic placing of the South Pole *in relation to Australia* in the Permian and Carboniferous is indicated (approximately) by A. In most cases the two poles are farther apart than would be expected if the continental reconstructions are correct.

For example, on du Toit's map of "Gondwana" (my Fig. 24) the Permo-Carboniferous glacial pole is slightly to the right of the eastern edge of Africa, and the Australian paleomagnetic pole is about 30° farther to the right (compass directions are ambiguous on a map like this). The discrepancy between the two pole positions is serious. If this reconstruction is correct, and if the Australian paleomagnetic pole was the true South Pole, then Permo-Carboniferous glaciation was very eccentric, and the farthest ice centers in southern Africa and South America were something like 60° of latitude from the pole. Such an eccentric pattern of glaciation can hardly be accepted without an explanation, and du Toit does not explain it.

A more or less similar discrepancy between the positions of the glacial and (Australian) paleomagnetic poles occurs on all the other hypothetical supercontinents tested (Figs. 25–28). On the more-open grouping of continents suggested in Fig. 38 there is still a discrepancy between the glacial and Australian paleomagnetic poles and the distribution of glaciated areas is still eccentric, but the eccentricity is explained (p. 181).

Glaciation, floras, and paleomagnetic evidence. The distributions of glaciation and of floras in the Permo-Carboniferous are related. Both are climatic indicators, not direct indicators of latitude. I think that, together, they prove that southern Africa, southern South America, southern Australia, and also India were much colder in the Permo-Carboniferous than they are now or than they were during the Pleistocene, and that the northern continents were warmer, and I think that the only acceptable explanation is that all the continents (except Antarctica) lay farther south then than now (Fig. 38). Paleomagnetism directly confirms a more southern position at this time only for North America, Europe, and Australia. However, the relatively good paleomagnetic placing of Australia not only confirms the position of that continent far southward at the time of glaciation but allows two important additional inferences about climate. These inferences are discussed in more detail in Chapter 14 and need only be summarized here. One is that, whether or not the continents have moved, forests existed and coal was formed in polar or subpolar regions in the Permo-Carboniferous. The finding of fossil forests and coal on Antarctica is proof only that the Antarctic climate was warmer than now, not that the continent occupied a different position. And the other inference, derived from comparison of the glacial and paleomagnetic records on Australia, is that the onset and ending of Permo-Carboniferous glaciation were due not to movements of land but to revolutions of climate comparable to the climatic revolution that brought on glaciation of northern continents in the Pleistocene.

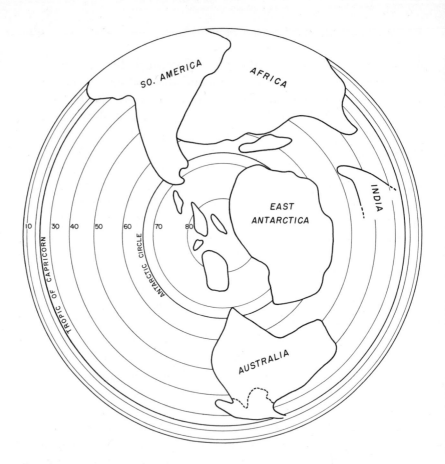

FIG. 38. Suggested arrangement of the continents in late Paleozoic time (reprinted from Darlington 1964:1086). West Antarctica, which is a relatively young, geologically active area, is shown as an archipelago, as it may still be, under the ice. New Zealand is omitted because I cannot determine its probable extent and position at this time (see Fleming 1962:58–59).

Northward movements of continents. The glacial and botanical evidence, confirmed in part by the incomplete record of paleomagnetism, strongly suggests that all the continents (except Antarctica) lay farther south on the world in the Permo-Carboniferous than they do now (preceding paragraph; and see again Fig. 38). If this is correct, most continents have moved northward since then, but the movements seem to have been uncoordinated. Africa and South America seem to have made moderate northward movements, and seem to have reached their present latitudes by the Jurassic (Creer 1958). Australia has made a moderate northward movement too but apparently did so later, for the paleomagnetic record shows this continent still far southward in the Jurassic (Fig. 37). And India seems to have moved northward farther and perhaps faster than Africa, South America, and Australia have done, mostly since the Jurassic (p. 112). Antarctica has apparently moved too, not northward but from a Jurassic position beside the pole (p. 115) to its present position over it. Some of these details may be doubtful, but the evidence does at least suggest that different pieces of land have moved, mostly northward, for different distances and at different times since the Permo-Carboniferous.

New Zealand existed in the Permo-Carboniferous but was apparently not glaciated then (p. 104). I therefore prefer not to suggest its position then or its movements (if any) later, except to note that its climatic record (Chapter 10) indicates a position south of the tropics through the whole time under consideration.

Ancient southern land bridges? If the southern continents and India were grouped far southward late in the Paleozoic (Fig. 38), were they connected by narrow land bridges—isthmian links—even if they were not broadly united? This question is less important than the question of continental movement and is more difficult to answer. Paleomagnetism, which may show latitudes and rotations but not shapes or connections of land, can tell nothing about southern land bridges. Geologic evidence requires none, so far as I know, and the distribution of Permo-Carboniferous glaciation seems to suggest the existence of an ocean current freely circling Antarctica, not blocked by land (p. 181). Biogeographic evidence probably does not require connections between southern continents. The *Glossopteris* flora may, I think, have dispersed across moderate ocean gaps (p. 193). And the fossil record of life on New Zealand (Fleming 1962; my Chapter 10) does not seem to show the making or breaking of any specific land connections, and suggests that none existed. These bits of evidence are very limited but do hint that not even narrow land bridges existed in the Southern Hemisphere during and after the Permo-Carboniferous.

21. Conclusions: history of land and life at the southern end of the world

Evidence from several independent sources, discussed and compared in preceding chapters, has forced me to conclude that the southern continents have drifted. I have therefore become a Wegenerian, but not an extreme one. I doubt the former existence of a Pangaea or Gondwanaland, and I think that movements of continents have been simpler and shorter than most Wegenerians suppose. Also, I am not absolutely sure of my conclusions. I shall therefore state the conclusions as probabilities rather than proven facts.

Geologic-geographic history. My tentative history of the southern end of the world begins late in the Paleozoic. All the continents (except Antarctica) probably lay farther south then than they do now. This is indicated by the distribution of glaciation and of floras in the Upper Carboniferous and Permian and by some paleomagnetic data, although enormous gaps still exist in the paleomagnetic record. South America may still have been joined to Africa at this time.

South America and Africa probably split apart (if they were joined) not later than the Jurassic, and if Creer's paleomagnetic inferences are correct, these two continents actually separated earlier than that, at the end of the Paleozoic or at latest before the Jurassic, and had reached approximately their present latitudes by the Jurassic. After that Africa, being near its present latitude and separate from South America, was hardly in position to play much part in the history of far-southern land and life.

Africa, South America, Australia, Antarctica, and India were apparently not broadly united (except for the hypothetical union of Africa and South America) but may have been grouped near the southern end of the world at the time of Permo-Carboniferous glaciation, as suggested in Fig. 38. Since then, different southern continents seem to have moved northward for different distances, at different times, and perhaps at different rates (Chapter 20), while

Antarctica seems to have moved toward the South Pole. The apparently independent movements of different continents seem incompatible with the existence of any southern supercontinent for any long period.

Attempted reconstructions of the positions and movements of southern continents since the late Paleozoic, based primarily on climatic indicators (distributions of glaciation and of floras) and on the incomplete paleomagnetic record, must satisfy an additional requirement, I think: all the now-habitable continents must always be reasonably accessible to terrestrial life, with continual opportunity for exchange between northern and southern land masses (Chapter 7).

Although some Wegenerians have postulated more complex movements, I think that the probabilities and requirements summarized in preceding paragraphs can best be met by keeping the continents as nearly as possible in their present relative positions and moving them no more than necessary. Africa, South America, Australia, and India may have moved northward in nearly straight lines from positions suggested in Fig. 38, to latitudes indicated by paleomagnetism in the mid-Mesozoic (Fig. 29), to present positions (Frontispiece) with no deviation and not much rotation. The northern continents have apparently moved northward too, from more-southern positions in the Permo-Carboniferous indicated by paleoclimatic and paleomagnetic evidence to their present positions. Their movements have apparently been correlated with the movements of the southern continents in such a way that no long interruptions have occurred in exchange of terrestrial life between northern and southern continents.

A general northward movement of continents, excepting Antarctica, *might* be correlated with the existing ridge-rift system of the earth. Apparent ridge-rift lines on the ocean bottom almost surround Antarctica (Wilson 1963a, b; Heezen and Ewing 1963; Girdler 1963) and *may* mark a line of rising and spreading convection currents in the earth's mantle that *may* have moved the other continents northward. I suggest this as a possibility to be considered, not as a probability that I am ready to defend. I am not technically qualified to defend it.

Whether or not narrow land connections have existed between southern continents in the remote past is relatively unimportant and very difficult to decide. Existing geologic and biogeographic

evidence apparently does not require connections in the late Paleozoic and Mesozoic and hints that none existed then (p. 209).
The record of mammals (Chapter 7) forbids southern land connections in the Tertiary, I think, except possibly for a temporary
connection between South America and Antarctica (p. 156).

Climatic history. The importance of climate in southern biogeography is stressed throughout this book. Local climatic factors have
been decisive in determining both the composition and the bound-
aries of each far-southern flora and fauna (Chapter 2). Zonation of
climate on the round earth has contributed both to the similarity
of far-southern biotas in different places and to their isolation from
the rest of the world. And in a more general way, by its influence
on world-wide patterns of evolution and dispersal, climate has (I
think) induced a continual flow of cold-tolerant plants and animals toward the southern end of the world (Chapters 5 and 6). The
prevalence of amphitropical distributions at different taxonomic
levels emphasizes that zonation of climate has been a fundamental factor in distribution of life in the far south for a very long time
(Chapter 14).

The earth's climate has probably always been zoned to some
extent. A zoneless climate on a round earth would be an anomaly
at any time. Moreover, the climatic zones have apparently been
oriented more or less as they are now, with respect to the main
pattern of land in the Southern Hemisphere, throughout the time
under consideration, since the late Paleozoic. Africa, South America, and Australia (but not India) now lie in such positions that the
areas glaciated in the Permo-Carboniferous are about equidistant
from the South Pole. This suggests that these three continents have
moved northward approximately equal distances since the late
Paleozoic (if they have moved at all) and that they then lay in a
zonal climate with zones oriented as now. Southern South America,
southern Australia, New Zealand, and Antarctica have all apparently been situated south of the tropics continuously since Permo-
Carboniferous glaciation. The evidence of this is given in Chapters
8, 9, 10, and 12, and is summarized in Chapter 13. The evidence is
consistent as far as it goes and indicates that, although the climate
of these pieces of land may have been (warm?) temperate for long
periods, it was cooler than tropical and probably seasonal at all
times for which an adequate record exists. I think biogeographers

should stop exaggerating the tropicalness of the southern end of the world in the past!

Although the climate at the southern end of the world has apparently been mild for long periods, two major cycles of cold occurred during the time under consideration. Both cycles were apparently due to revolutions of climate, not (as some Wegenerians have thought) to movements of continents across the south polar region. One climatic revolution occurred in the late Carboniferous and early Permian and caused heavy glaciation on southern continents. The extent of the ice sheets suggests that the glaciated continents lay far southward then, but the onset and ending of glaciation were apparently due to changes of climate rather than to movement of land (p. 134). The second revolution of climate culminated in Pleistocene glaciation, which was relatively light on southern South America and Australia and absent on South Africa (suggesting that these lands now lie farther from the South Pole than they did before) but which buried Antarctica under ice. The occurrence of the Pleistocene climatic revolution parallels and helps to substantiate the Permo-Carboniferous one. The causes of the two may have been the same. But, of course, the causes even of the more recent glacial cycle are still doubtful.

Biotic history. In the late Carboniferous and early Permian, during short interglacial periods or after glaciation or perhaps actually at the edges of the ice sheets, the *Glossopteris* flora apparently covered all far-southern glaciated areas. It was a relatively species-poor, specialized flora, apparently adapted primarily to glacial and postglacial conditions. It was remarkably uniform on the different glaciated continents, with a high proportion of apparently identical species even on Antarctica and India (p. 111). The northern limits of this flora were (I think) probably determined by warmer climate. The flora may have had no southern limit. This and perhaps some later far-southern floras apparently extended into south polar regions and formed coal there (p. 133).

Even after disappearance of Permo-Carboniferous ice, during nonglacial geologic periods, far-southern terrestrial life has probably always had to be adapted to a relatively cool, seasonal climate. And climatic barriers have probably always limited the distribution of far-southern groups toward the north and retarded exchange of plants and animals with warmer regions, accentuating biotic relationships between different southern land areas. These areas, including at least parts of Antarctica, have probably shared characteristic forests of gymnosperms, including conifers, during much of the Mesozoic and of angiosperms, including *Nothofagus,* later, during the late Cretaceous and Tertiary. And characteristic groups of invertebrates have probably always inhabited the forests. Moorland and its invertebrates have probably been widespread too.

If my hypothetical plan (Fig. 38) is approximately correct, dispersal of far-southern plants and animals has been easier in the past than now. Far-southern groups presumably dispersed across relatively narrow water gaps via Antarctica when the latter was habitable. Dispersal may have been across actual southern isthmian land connections during the late Paleozoic and early Mesozoic but (I think) connections are not required and not indicated, and in any case dispersal probably continued across the water gaps as long as even the edges of Antarctica were habitable.

The biogeographic history of New Zealand (Chapter 10) is especially significant in this connection and is worth resummarizing. Fleming (1962; 1964), in his informative reviews, stresses the continual interplay of cool southern and tropical or subtropical Indo-Pacific elements in the fossil record on New Zealand and the apparent continual arrivals of new forms from several different directions. The record seems to show no sudden massive influxes of plants and animals from one direction and no sudden cessations of dispersal such as might be expected to have accompanied the making and breaking of land connections, and other evidence seems to indicate long isolation of New Zealand. I think, therefore, that New Zealand has probably been isolated since the Carboniferous (or longer) and has received its life across deep ocean barriers. And, because New Zealand has received a large proportion of the groups of terrestrial organisms characteristic of other far-southern lands, including southern conifers (prominent in New Zealand vegetation at least since the Jurassic—Fleming 1962:104), many angiosperms, and many invertebrates (but relatively few terrestrial vertebrates), I think that these plants and animals crossed ocean gaps to New Zealand and that many other plants and invertebrates of the same general groups may have crossed comparable water gaps elsewhere in the Southern Hemisphere.

Summary of conclusions. I think, then, that at the time of the Permo-Carboniferous glaciation all the continents (except Antarctica) probably lay farther south than now. Africa and South America may still have been joined together but the other southern continents were probably not united. Africa and South America separated not later than the Jurassic and probably earlier, too long ago for their union to affect the distribution of plants and animals now. And all the continents (except Antarctica) have moved northward since the Permo-Carboniferous, different continents moving independently. Water gaps have probably been narrower and dispersal has probably been easier in the past than now, but land bridges between southern continents are not indicated.

Even when the land was not glaciated, the climate of the southern end of the world since the Permo-Carboniferous has probably always been cooler than tropical and also seasonal, and zonation of climate has probably always favored differentiation of a special southern cool-temperate biota. The biotic history of the southern end of the world seems in fact to have been the history of a continuously existing but continually changing, climatically specialized, far-southern biota shared at least in part by all far-southern lands, including habitable parts of Antarctica. Beginning with near-contiguity of southern continents in the late Paleozoic, facilitating dispersal in the far south, the history has probably been one of gradual widening of gaps between continents and gradual lessening of dispersal and finally deterioration of climate and virtual cessation of dispersal across southern water gaps during the later Tertiary and Pleistocene.

During the whole of the time in question, successive new groups of plants and animals have presumably been invading the southern end of the world from the tropics or across the tropics, and minor counterinvasions from the south northward may have occurred too. However, I see no evidence that any major groups of plants or animals have evolved on Antarctica and spread widely over the world from there (p. 153). During this whole time dispersal around the southern end of the world has probably tended to be from west to east, since that has probably always been the direction of prevailing winds and currents.

The tidal-river analogy ties all this together. The biotic tide has (I think) probably been running irregularly but strongly toward the southern end of the world since the Paleozoic, perhaps by several routes, with some countercurrents and endless minor complexities in the main flow; and the main currents have been feeding into a gigantic whirlpool of dispersal moving mainly from west to east around the southern end of the world, while the far-southern biota has been continually changing by arrival of new forms, extinction of old ones, and evolution in the far south. This whole process is (I think) clearly visible in records of the more recent past, less visible in the more remote past, but it is based on general considerations (Chapters 5 and 6) as well as on specific evidence. It is based also on the principle that a dispersal pattern that exists now and that seems to be a product of an existing situation is likely to be as old as the situation that produces it. The essential features of the situation are zonation of climate, limitation of land areas south of the tropics, and existence of a system of powerful winds and ocean currents circling the southern end of the world.

In reaching these conclusions about the history of land and life at the southern end of the world, I have oversimplified some problems and some explanations. I regret the necessity of this. I do not trust Occam's razor. The simplest explanations are not necessarily the right ones in biogeography. To choose the simplest explanation because it is simple is like a surgeon choosing to cut a patient's throat with one razor stroke rather than to perform a complex operation. Occam's razor should be used to make an exploratory cut into a problem, not to solve it. Nevertheless I have oversimplified, because I have had to, and in doing so I may unintentionally have cut out essential facts or interpretations. Have I?

POSTSCRIPT

This postscript (March 26, 1965) brings together important new information that has come to hand while my book was in press.

History of Antarctica (Chapter 12). Paleomagnetic determinations made by D. J. Blundell and summarized by R. J. Adie (in Priestley *et al.* 1964:154) show the Antarctic Peninsula (Graham Land) near its present latitude during the Tertiary (when forest including *Nothofagus* occurred on the Peninsula) but place the tip of the Peninsula at 35°–40° S earlier than that (when parts of Antarctica were more heavily forested). A latitude of 35°–40° is equivalent to the present latitude of central Chile or (in the Northern Hemisphere) central California.

The Scotia Arc, which lies between the Antarctic Peninsula and the tip of South America (broken line in Frontispiece), apparently includes sediments as old as the Carboniferous (Adie, in Priestley *et al.*, p. 119). This seems to imply that Antarctica and South America were separate then.

D. L. Linton (in Priestley *et al.*, pp. 87–88, Fig. 1) shows that the climate of the western side of the Antarctic Peninsula is now milder and more oceanic than the climate of the mainland of Antarctica. I have stressed the probable importance of this difference in the past.

Collembola and mites have now been found near Shackleton Glacier on Antarctica, extending the known southern limit of terrestrial animals to 84° 35′ S, within 640 km of the South Pole (Wise 1964).

History of Nothofagus (Chapter 16). Pollen of *Nothofagus,* and of other now-southern plants, is reported in southern Eurasia, Kazakhstan, and western Siberia in Upper Cretaceous and early Tertiary deposits (E. D. Zaklinskaya, in Cranwell 1964b:85).

In southeastern Australia, *Nothofagus* pollen has not been found in older deposits that have been examined but appears in the Paleocene to Lower Eocene (I. C. Cookson, in Cranwell 1964b:81), suggesting (I think) an actual arrival then. Pollen of the *brassii* type seems to have appeared first and was the commonest type of *Nothofagus* pollen in Australia in the Tertiary.

Paleomagnetic evidence (Chapter 20). Additional reading and correspondence convince me that the limits of paleomagnetic evidence should be further emphasized. Paleomagnetic determinations of latitude (distance from a pole) are based on well-supported assumptions and indicate great changes of latitude of individual continents. Paleomagnetic determinations of longitude (relative east-west position) depend on an *additional* assumption, that the poles have moved or "wandered" at times when individual continents did not move (Irving 1964:273). This assumption is not supported by convincing evidence (Deutsch 1963:4). It is doubtful if ancient longitudes can ever be found by comparison of polar-wandering curves, and good paleomagnetic determinations from the southern continents are still too few even to test the method satisfactorily. Also I now find that the "preliminary" determinations from which Creer (1958) calculated divergence of Africa and South America in the Jurassic are not trustworthy and can be interpreted in other ways.

I therefore now think that supposed paleomagnetic indications of the time of separation of Africa and South America (Chapter 18) and of break-up of "Gondwanaland" (Chapter 20) are valueless and should be entirely disregarded.

The first technically satisfactory paleomagnetic determinations for Africa near the time of Permo-Carboniferous glaciation (Opdyke 1964:2477–2487, Fig. 12) give Africa almost exactly the latitude and orientation suggested in my Fig. 38.

Bibliography

The subjects included in this reference list are diverse and much has been published on some of them recently. Therefore, in order to keep the whole list within reasonable limits, I have selected a small number of references from among the many available on some subjects. My selection has probably been capricious in some cases, but I have had to select, and I can only offer my regrets and apologies to the many persons whose published work has been omitted. The items listed below are principally those actually referred to in the preceding pages. However, for the information of interested readers, I have included a few recent books that are not referred to elsewhere and some of which I have not seen. Some are in fact still in press.

Adie, R. J.
  1965   *Antarctic geology* (Wiley, Interscience, New York).

Ager, D. V.
  1963   *Principles of paleoecology. An introduction to the study of how and where animals and plants lived in the past* (McGraw-Hill, New York).

American Geographical Society (New York)
  1956   Map of *Tierra del Fuego,* 1 : 1,000,000.
  1962   Map of *Antarctica,* 1 : 5,000,000.
  1964–   *Antarctic map folio series.*

Athavale, R. N., C. Radhakrishnumurty, and P. W. Sahasrabudhe
  1963   "Palaeomagnetism of some Indian rocks," *Geophysical J.* (Royal Astronomical Soc.) *7*:304–313.

Auer, V.
  1960   "The Quaternary history of Fuego-Patagonia," in Pantin 1960:507–516.

Australian Encyclopaedia, The
  1958   Ten vols. (Halstead Press, Sydney, Australia; Michigan State Univ. Press, E. Lansing, Mich.).

Ball, G. E.
  1956   "Notes . . . on the classification of the tribe Broscini . . ." *Coleopterists' Bull. 10*:33–52.

Banks, M. R.
  1962   "Permian," in Spry and Banks 1962:189–215.

Barber, H. N., H. E. Dadswell, and H. D. Ingle
  1959   "Transport of driftwood from South America to Tasmania and Macquarie Island," *Nature 184*:203–204.

Barghoorn, E. S.
    1961 "A brief review of fossil plants of Antarctica and their geologic implications," in Nat. Acad. Sci.—Nat. Research Council 1961, Part 1:5–9.

Behrendt, J. C., and P. E. Parks, Jr.
    1962 "Antarctic Peninsula traverse," *Science 137*:601–602.

Blackett, P. M. S., *et al.* (ed.)
    1965 "Continental drift," *Proc. Royal Soc. (London)* (in press).

Brink, P.
    1957 "Onychophora," in *South African Animal Life* (Results Lund Univ. Expedition 1950–1951; Almquist and Wiksells, Uppsala), Vol. 4:7–32.
    1960 "The relation between the South African fauna and the terrestrial and limnic animal life of the southern cold temperate zone," in Pantin 1960:568–571.

Britton, E. B.
    1949 "The Carabidae (Coleoptera) of New Zealand. Part III—A revision of the tribe Broscini," *Trans. Royal Soc. New Zealand 77*:533–581.
    1959 "Carabidae (Coleoptera) from New Zealand caves," *Proc. Royal Entomol. Soc. London (B) 28*:103–106.
    1962 "New genera of beetles (Carabidae) from New Zealand," *Ann. Mag. Nat. Hist.* (13) *4*:665–672.

Brooks, C. E. P.
    1949 *Climate through the ages* (2nd ed., McGraw-Hill, New York).

Bucher, W.
    1962 Review of Nairn, *Descriptive paleoclimatology, Am. Scientist 50*:296A–300A.

Burbidge, N. T.
    1960 "The phytogeography of the Australian Region," *Australian J. Botany 8*:75–211.

Carey, S. W. (ed.)
    1958 *Continental drift. A symposium* (Geology Department, Univ. of Tasmania, Hobart, Tasmania).

Carrick, R., M. Holdgate, and J. Prevost (ed.)
    1964 *Biologie antarctique* (Hermann, Paris).

Caster, K. E.
    1952 "Stratigraphic and paleontologic data relevant to the problem of Afro-American ligation during the Paleozoic and Mesozoic," in Mayr 1952:105–152.

Caughley, G.
    1964 "Does the New Zealand vertebrate fauna conform to zoogeographic principles?" *Tuatara* (J. Biol. Soc. Victoria Univ., Wellington, N. Z.) *12*:49–56.

Childs, O. E., and B. W. Beebe (ed.)
    1963 *Backbone of the Americas. A symposium. Tectonic history from pole to pole* (Am. Assoc. Petroleum Geol., Memoir 2, Tulsa, Okla.).

China, W. E.
    1962 "South American Peloridiidae . . ." *Trans. Royal Entomol. Soc. London 114*:131–161.

Cloud, P. E., Jr.
    1961 "Paleobiogeography of the marine realm," in Sears 1961:151–200.

Cockayne, L.
    1928 *The vegetation of New Zealand* (2nd ed., Engelmann, Leipsig).

Colbert, E. H.
    1952   "The Mesozoic tetrapods of South America," in Mayr 1952:237–249.
    1961   *Dinosaurs: their discovery and their world* (Dutton, New York).

Coleman, A. P.
    1926   *Ice ages recent and ancient* (Macmillan, New York).

CSIRO [Commonwealth Scientific and Industrial Research Organization].
    1950   *The Australian environment* (2nd ed., CSIRO, Melbourne, Australia).

Constance, L. (ed.).
    1963   "Amphitropical relationships in the herbaceous flora of the Pacific
            Coast of North and South America: a symposium," *Quart. Rev. Biol.*
            38:109–177.

Corner, E. J. H.
    1964   "*Ficus* in the Pacific Region," in Gressitt 1964d:233–245.

Couper, R. A.
    1960   "Southern Hemisphere Mesozoic and Tertiary Podocarpaceae and
            Fagaceae and their palaeogeographic significance," in Pantin 1960:491–
            500.

Cox, A., and R. R. Doell
    1960   "Review of paleomagnetism," *Bull. Geol. Soc. Am. 71*:645–768.

Craig, G. Y.
    1961   "Palaeozoological evidence of climate. (2) Invertebrates," in Nairn
            1961:207–226.

Cranwell, L. M.
    1964a   "*Nothofagus*: living and fossil," in Gressitt 1964d:387–400.
    1964b   (Ed.) *Ancient Pacific floras: the pollen story* (Univ. of Hawaii Press, Hon-
            olulu).

Cranwell, L. M., H. J. Harrington, and I. G. Speden
    1960   "Lower Tertiary microfossils from McMurdo Sound, Antarctica,"
            Nature 186:700–702.

Crary, A. P. (ed.)
    1956   *Antarctica in the International Geophysical Year* (Nat. Acad. Sci., Washing-
            ton, Publ. 462).

Crawshay, R.
    1907   *The birds of Tierra del Fuego* (Quaritch, London).

Creer, K. M.
    1958   "Preliminary palaeomagnetic measurements from South America,"
            Ann. Géophysique 14:373–390.
    1964a   "A reconstruction of the continents for the Upper Palaeozoic from
            palaeomagnetic data," *Nature 203*:1115–1120.
    1964b   "Palaeomagnetic data and du Toit's reconstruction of Gondwana-
            land," *Nature 204*:369–370.

Darlington, P. J., Jr.
    1936   "Variation and atrophy of flying wings of some carabid beetles," *Ann.
            Entomol. Soc. Am. 29*:136–176.
    1938a   "The origin of the fauna of the Greater Antilles, with discussion of
            dispersal of animals over water and through the air," *Quart. Rev. Biol.
            13*:274–300.
    1938b   "Was there an Archatlantis?" *Am. Naturalist 72*:521–533.
    1943   "Carabidae of mountains and islands: data on the evolution of isolated
            faunas, and on atrophy of wings," *Ecol. Monographs 13*:37–61.

1948 "The geographical distribution of cold-blooded vertebrates," *Quart. Rev. Biol. 23*:1–26, 105–123.

1949 "Beetles and continents" (Review of Jeannel, *La genèse des faunes terrestres*), *Quart. Rev. Biol. 24*:342–345.

1953 "A new *Bembidion* (Carabidae) of zoogeographic importance from the Southwest Pacific," *Coleopterists' Bull. 7*:12–16

1957 *Zoogeography: the geographical distribution of animals* (Wiley, New York).

1959a "Area, climate, and evolution," *Evolution 13*:488–510.

1959b "Darwin and zoogeography," *Proc. Am. Phil. Soc. 103*:307–319.

1959c "The *Bembidion* and *Trechus* (Col.: Carabidae) of the Malay Archipelago," *Pacific Insects* (Bishop Mus., Honolulu) *1*:331–344.

1960a "The zoogeography of the southern cold temperate zone," in Pantin 1960:659–668.

1960b "Australian carabid beetles IV. List of localities, 1956–1958," *Psyche 67*:111–126.

1961a "Australian carabid beetles V. Transition of wet forest faunas from New Guinea to Tasmania," *Psyche 68*:1–24.

1961b "Australian carabid beetles VII. *Trichosternus,* especially the tropical species," *Psyche 68*:113–130.

1961c "Australian carabid beetles IX. The tropical *Notonomus,*" *Breviora* (Mus. Comp. Zool.), No. 148:1–14.

1962 "Australian carabid beetles X. *Bembidion,*" *Breviora* (Mus. Comp. Zool.), No. 162:1–12.

1964 "Drifting continents and late Paleozoic geography," *Proc. Nat. Acad. Sci. 52*:1084–1091.

Darwin, C.
1859 *On the origin of species* (John Murray, London; facsimile of 1st ed., Harvard Univ. Press, Cambridge, Mass., 1964).

David, T. W. E.
1950 *The geology of the Commonwealth of Australia* (Arnold, London).

Deutsch, E. R.
1963 "Polar wandering and continental drift: an evaluation of recent evidence," in Munyan 1963:4–46.

Dicke, R. H.
1962 "The earth and cosmology," *Science 138*:653–664.

Dietz, R. S.
1964 "The third surface," in Miller 1964:3–15.

Dobzhansky, T.
1950 "Evolution in the tropics," *Am. Scientist 38*:209–221.

Doumani, G. A., and W. E. Long
1962 "The ancient life of the Antarctic," *Sci. Am. 207,* No. 3 (Antarctic number, Sept.), pp. 169–184.

Doumani, G. A., and P. Tasch
1963 "Laeiid conchostracan zone in Antarctica and its Gondwana equivalents," *Science 142*:591–592.

Durham, J. W.
1964 "Paleogeographic conclusions in light of biological data," in Gressitt 1964d:355–365.

du Toit, A. L.
1927 *A geological comparison of South America with South Africa* (Carnegie Inst., Washington, Publ. 381).

1937   *Our wandering continents: an hypothesis of continental drift* (Oliver & Boyd, Edinburgh).
1954   *Geology of South Africa* (3rd ed., Oliver & Boyd, Edinburgh).

Eigenmann, C. H.
1909   "The fresh-water fishes of Patagonia and an examination of the Archiplata-Archhelenis theory," Rep. Princeton Univ. Expedition Patagonia, *1896-1899*, Vol. 3, Part 3:225-374.

Euller, J.
1960   *Antarctic world* (Abelard-Schuman, New York).

Evans, J. W.
1959   "The Peloridiidae of Lord Howe Island . . ." *Records Australian Mus.* *25*:57-62.

Faegri, K., and J. Iversen
1964   *Textbook of pollen analysis* (2nd ed., Hafner, New York).

Fisher, O.
1882   "On the physical cause of the ocean basins," *Nature 25*:243-244.

Fleming, C. A.
1962   "New Zealand biogeography. A paleontologist's approach," *Tuatara* (J. Biol. Soc. Victoria Univ., Wellington, N. Z.) *10*:53-108.
1964   "Paleontology and southern biogeography," in Gressitt 1964d:369-385.

Flint, R. F.
1957   *Glacial and Pleistocene geology* (Wiley, New York).

Florin, R.
1963   "The distribution of conifer and taxad genera in time and space," *Acta Horti Bergiani* (Almquist & Wiksells, Uppsala) *20*:121-312.

Furness, F. N. (ed.)
1961   *Solar variation, climatic change, and related geophysical problems* (New York Acad. Sci., New York, Annals Vol. 95).

Gentilli, J.
1961   "Quaternary climates of the Australian Region," in Furness 1961:465-501.

Gill, E. D.
1961   "The climates of Gondwanaland in Kainozoic times," in Nairn 1961:332-353.
1962   "Cainozoic [of Tasmania]," in Spry and Banks 1962:233-254.

Girdler, R. W.
1963   "Rift valleys, continental drift and convection in the earth's mantle," *Nature 198*:1037-1039.

Gollerbakh, M. M., and E. E. Syroechkovskyi
1960   [Biogeographic explorations in the Antarctic], *Biol. Abstr. 45*, Feb. 1964, No. 13,265.

Gressitt, J. L.
1964a   "Insects of Antarctica and subantarctic islands," in Gressitt 1964d:435-442.
1964b   "Ecology and biogeography of land arthropods in Antarctica," in Carrick 1964:211-222.
1964c   "Summary," in Gressitt *et al.*, *Insects of Campbell Island* (Pacific Insects Monograph 7, Bishop Mus., Honolulu), pp. 531-600.
1964d   (Ed.) *Pacific Basin biogeography* (Bishop Mus. Press. Honolulu [dated 1963]).

Gressitt, J. L., R. E. Leech, and K. A. J. Wise
    1963 "Entomological investigations in Antarctica," *Pacific Insects* (Bishop Mus., Honolulu) 5:287–304.

Gressitt, J. L., and C. M. Yoshimoto
    1964 "Dispersal of animals in the Pacific," in Gressitt 1964d:283–292.

Guppy, H. B.
    1906 *Observations of a naturalist in the Pacific between 1896 and 1899. II. Plant dispersal* (Macmillan, London).
    1917 *Plants, seeds and currents in the West Indies and Azores* (Williams & Norgate, London).

Hafsten, U.
    1951 "A pollen-analytical investigation of two peat deposits from Tristan da Cunha," *Results Norwegian Sci. Expedition Tristan da Cunha 1937–1938,* No. 22:1–43.

Hamilton, T. H., R. H. Barth, Jr., and I. Rubinoff
    1964 "The environmental control of insular variation in bird species abundance," *Proc. Nat. Acad. Sci. 52*:132–140.

Hamilton, W.
    1963 "Tectonics of Antarctica," in Childs and Beebe 1963:4–15.

Harrington, H. J.
    1962 "Paleogeographic development of South America," *Bull. Am. Assoc. Petroleum Geol. 46*:1773–1814.

Haughton, S. H.
    1952 "The Karroo System in Union of South Africa," *XIX Congrès Géol. Internat. Symposium Series de Gondwana,* pp. 254–255.
    1963 *The stratigraphic history of Africa south of the Sahara* (Oliver & Boyd, Edinburgh).

Heezen, B. C., and M. Ewing
    1963 "The mid-ocean ridge," in Hill 1963:388–410.

Hill, M. N. (ed.)
    1963 *The sea. Vol. 3. The earth beneath the sea. History* (Wiley, New York).

Holdgate, M. W.
    1964 Discussion in Carrick 1964:246.

Holloway, J. T.
    1954 "Forests and climates in the South Island of New Zealand," *Trans. Royal Soc. New Zealand 82*:329–410.

Hooker, J. D. *See* Turrill 1953.

Imbrie, J., and N. D. Newell (ed.)
    1964 *Approaches to paleoecology* (Wiley, New York).

Irving, E.
    1964 *Paleomagnetism and its application to geological and geophysical problems* (Wiley, New York).

Irving, E., and D. A. Brown
    1964 "Abundance and diversity of the labyrinthodonts as a function of paleolatitude," *Am. J. Sci. 262*:689–708.

Jeannel, R.
    1926–1928. "Monographie des Trechinae," *L'Abeille 32,* No. 3; *33; 35.*
    1938 "Les migadopides," *Rev. Française d'Entomol. 5*:1–55.

1941   *Coléoptères carabiques, première partie,* vol. 39 of *Faune de France* (Lechevalier, Paris).

1942   *La genèse des faunes terrestres* . . . (Presses Universitaires de France, Paris).

1962   "Les trechides de la Paléantarctide Occidentale," *Biologie de l'Amérique Australe* (Centre Nat. Recherche Sci., 15 Quai Anatole-France, Paris), vol. 1, *Etudes sur la faune du sol,* pp. 527–655.

Keast, A.
1961   "Bird speciation on the Australian continent," *Bull. Mus. Comp. Zool. 123*:305–495.

King, L. C.
1961   "The palaeoclimatology of Gondwanaland during the Palaeozoic and Mesozoic Eras," in Nairn 1961:307–331

1962   *The morphology of the earth. A study and synthesis of world scenery* (Hafner, New York).

Kort, V. G.
1962   "The Antarctic Ocean," *Sci. Am. 207,* No. 3 (Antarctic number, Sept.), pp. 113–128.

Kräusel, R.
1961   "Palaeobotanical evidence of climate," in Nairn 1961:227–254.

Kummel, B.
1961   *History of the earth: an introduction to historical geology* (Freeman, San Francisco).

Kuschel, G.
1960   "Terrestrial zoology in southern Chile," in Pantin 1960:540–550.

1964   "Problems concerning an Austral Region," in Gressitt 1964d:443–449.

Kusnezov, N.
1957   "Numbers of species of ants in faunae of different latitudes," *Evolution 11*:298–299.

Landsberg, H. E., H. Lippmann, K. H. Paffen, and C. Troll
1963   *World maps of climatology* (Springer-Verlag, Berlin).

Levyns, M. R.
1962   "Possible antarctic elements in the South African flora," *South African J. Sci. 58*:237–241.

Lindroth, C. H.
1946   "Inheritance of wing dimorphism in *Pterostichus anthracinus* Ill.," *Hereditas 32*:37–40.

1949   *Die fennoskandischen Carabidae. Eine tiergeographische Studie. III. Allgemeiner Teil* (Göteborgs Mus. Zool., Stockholm).

1963a   *The ground-beetles of Canada and Alaska,* Parts 2 and 3 (*Opuscula Entomol.,* Suppl. 20 and 24, Lund, Sweden).

1963b   *The fauna history of Newfoundland illustrated by carabid beetles* (*Opuscula Entomol.,* Suppl. 23).

Llano, G. A.
1962   "The terrestrial life of the Antarctic," *Sci. Am. 207,* No. 3 (Antarctic number, Sept.), pp. 212–230.

Lord, C. E., and H. H. Scott
1924   *A synopsis of the vertebrate animals of Tasmania* (Oldham, Beddome, and Meredith, Hobart, Tasmania).

McDowall, R. M.
  1964 "The affinities and derivation of the New Zealand fresh-water fish fauna," *Tuatara* (J. Biol. Soc. Victoria Univ. Wellington, N. Z.) *12*:59–67.

Mason, R.
  1961 "Dispersal of tropical seeds by ocean currents," *Nature 191*:408–409.

Matthew, W. D.
  1915 *Climate and evolution (Ann. New York Acad. Sci. 24*:171–318; reprinted (1939) as Special Publ. New York Acad. Sci., 1).

Mayr, E.
  1952 (Ed.) "The problem of land connections across the South Atlantic . . ." *Bull. Am. Mus. Nat. Hist.* 99:79–258.
  1963 *Animal species and evolution* (Harvard Univ. Press, Cambridge, Mass.).

Meinesz, F. A. V.
  1964 *The earth's crust and mantle* (Elsevier, New York).

Menard, H. W., and H. S. Ladd
  1963 "Oceanic islands, seamounts, guyots and atolls," in Hill 1963:365–387.

Miller, R. L. (ed.)
  1964 *Papers in marine geology: Shepard commemorative volume* (Macmillan, New York).

Munyan, A. C. (ed.)
  1963 *Polar wandering and continental drift* (Soc. Economic Paleontologists and Mineralogists, Tulsa, Okla., Special Publ. 10).

Myers, G. S.
  1938 "Fresh-water fishes and West Indian zoogeography," *Smithsonian Rep.* for 1937, pp. 339–364.

Nairn, A. E. M. (ed.)
  1961 *Descriptive palaeoclimatology* (Interscience Publishers, New York).
  1964 *Problems in palaeoclimatology* (Wiley, New York).

National Academy of Sciences–National Research Council (Washington)
  1961 *Science in Antarctica. Part I, The life sciences; Part II, The physical sciences* (Publ. No. 839; 878).

Netolitzky, F.
  1942–1943 "Bestimmungstabellen . . . Gattung *Bembidion* Latr.," *Koleopterologische Rundschau,* Band 28, Heft 1/3, 3/6; Band 29, Heft 1/3.

Olson, E. C.
  1955 "Parallelism in the evolution of the Permian reptile faunas of the Old and New Worlds," *Fieldiana* (Chicago Nat. Hist. Mus.). *Zool. 37*:385–401.

Opdyke, N. D.
  1964 "The paleomagnetism of the Permian Red Beds of Southwest Tanganyika," *J. Geophysical Research 69*:2477–2487.

Osgood, W. H.
  1943 *The mammals of Chile* (Field Mus., Chicago, Zool. Series, 30).

Palaeogeography, climatology, ecology
  1965– A new journal, to be published by Elsevier, New York.

Pantin, C. F. A. (ed.)
  1960 "A discussion on the biology of the southern cold temperate zone," *Proc. Royal Soc. (London), Ser. B, 152*:429–677.

Parasnis, D. S.
    1961   *Magnetism* (Harper, New York).
Plumstead, E. P.
    1956   "Bisexual fructifications borne on *Glossopteris* leaves from South Africa,"
           Palaeontographica 100, Abt. B:1–25.
    1962   "Fossil floras of Antarctica," *Trans-Antarctic Expedition 1955–1958, Sci.
           Rep. 9*:1–154.
Preest, D. S.
    1964   "A note on the dispersal characteristics of the seed of the New Zealand
           podocarps and beeches and their biogeographical significance," in
           Gressitt 1964d:415–424.
Priestley, R., R. J. Adie, and G. de Q. Robin (ed.)
    1964   *Antarctic research* (Butterworths, London).
Rausch, R. L.
    1964   "A review of the distribution of Holarctic Recent mammals," in Gres-
           sitt 1964d:29–43.
Richards, P. W.
    1952   *The tropical rain forest. An ecological study* (Cambridge Univ. Press, Cam-
           bridge, England).
Robbins, R. G.
    1961   "The montane vegetation of New Guinea," *Tuatara* (J. Biol. Soc. Vic-
           toria Univ., Wellington, N. Z.) *8*:121–133.
Robin, G. de Q.
    1964   "Soviet antarctic research," *Nature 204*:110.
Romer, A. S.
    1945   *Vertebrate paleontology* (2nd ed., Univ. of Chicago Press, Chicago).
    1952   "Discussion [of Colbert 1952]," in Mayr 1952:250–254.
Rubin, M. J.
    1962   "The Antarctic and the weather," *Sci. Am. 207*, No. 3 (Antarctic num-
           ber, Sept.), pp. 84–94.
Runcorn, S. K. (ed.)
    1962   *Continental drift* (Academic Press, New York).
Schwarzbach, M.
    1963   *Climates of the past* (Van Nostrand, London).
Scientific American
    1962   *207*, No. 3 (Sept.), "The Antarctic."
Scott, W. B.
    1937   *A history of land mammals in the Western Hemisphere* (rev. ed., Macmillan,
           New York).
Sears, M. (ed.)
    1961   *Oceanography* (Am. Assoc. Advancement Sci., Washington, Publ. 67).
Serié, P.
    1936   ". . . Distributión geográfica de los ofidios argentinos," *Obra Cincuen-
           tenario Mus. de La Plata 2*:33–61.
Seward, A. C.
    1931   *Plant life through the ages* (Macmillan, New York).
Simpson, G. G.
    1940a "Antarctica as a faunal migration route," *Proc. Sixth Pacific Sci. Congr.
           2*:755–768.

1940b "Mammals and land bridges," *J. Washington Acad. Sci. 30*:137–163.
1945   *The principles of classification and a classification of mammals* (Am. Mus. Nat. Hist., New York, Bull. 85).
1947   "Evolution, interchange, and resemblance of the North American and Eurasian Cenozoic mammalian faunas," *Evolution 1*:218–220.
1950   "History of the fauna of Latin America," *Am. Scientist 38*:361–389.
1961   "Historical zoogeography of Australian mammals," *Evolution 15*:431–446.

Sloane, T. G.
1906   "Revision of the Cicindelidae of Australia," *Proc. Linnean Soc. New South Wales 31*:309–360.
1920   "The Carabidae of Tasmania," *Proc. Linnean Soc. New South Wales 45*:113–178.

Snider, A.
1859   *La création et ses mystères dévoilés* (Paris).

Solomon, M.
1962   "The tectonic history of Tasmania," in Spry and Banks 1962:311–339.

Soviet Antarctic Expedition
1964   *Information Bulletin* (Elsevier, New York); and see *Nature 205* (1965):235.

Spry, A., and M. R. Banks (ed.)
1962   *The geology of Tasmania* (*J. Geol. Soc. Australia* (Adelaide) 9, part 2).

Steenis, C. G. G. J. van. *See* van Steenis

Stehli, F. G., and C. E. Helsley
1963   "Paleontologic technique for defining ancient pole positions," *Science 142*:1057–1059.

Stuckenberg, B. R.
1962   "The distribution of the montane palaeogenic elements in the South African invertebrate fauna," *Ann. Cape Provincial Mus. 2*:190–205.

Sturgis, B. B.
1928   *Field book of birds of the Panama Canal Zone* (Putnam, New York).

Thorson, G.
1961   "Length of pelagic larval life in marine bottom invertebrates as related to larval transport by ocean currents," in Sears 1961:455–474.

Todd, W. E. C., and M. A. Carriker, Jr.
1922   *The birds of the Santa Marta Region of Colombia* (Carnegie Mus., Pittsburgh, Ann. 14).

Troll, C.
1960   "The relationship between the climates, ecology and plant geography of the southern cold temperate zone and of the tropical high mountains," in Pantin 1960:529–532.

Turnbull, G.
1959   "Some palaeomagnetic measurements in Antarctica," *Arctic 12*:151–157.

Turrill, W. B.
1953   *Pioneer plant geography: the phytogeographical researches of Sir Joseph Dalton Hooker* (Martinus Nijhoff, The Hague).

Umbgrove, J. H. F.
1949   *Structural history of the East Indies* (Cambridge Univ. Press, Cambridge, England).

van Oye, P., and J. van Mieghem (ed.)
  1964   *Biogeography and ecology of Antarctica* (W. Junk, The Hague, Monographiae Biologicae 15).

van Steenis, C. G. G. J.
  1953   "Results of the Archbold Expeditions. Papuan *Nothofagus*," *J. Arnold Arboretum 34*:301–373.
  1962   "The land-bridge theory in botany," *Blumea* (Rijksherbarium, Leiden, Netherlands) *11*:235–372.

Verdoorn, F., et al.
  1945   *Plants and plant science in Latin America* (Chronica Botanica, Waltham, Mass.).

von Ihering, H.
  1900   "The history of the Neotropical Region," *Science 12*:857–864.

Wegener, A.
  1915   *Die Entstehung der Kontinente und Ozeane* (Sammlung Vieweg, Brunswick; No. 23).
  1937   *La genèse des continents et des océans* (Librairie Nizet et Bastard, Paris).

Wexler, H., M. J. Rubin, and J. E. Caskey (ed.)
  1962   *Antarctic research* (Nat. Acad. Sci., Washington, Publ. 1036).

Wilson, E. O.
  1961   "Nature of the taxon cycle in the Melanesian ant fauna," *Am. Naturalist 95*:169–193.

Wilson, J. T.
  1963a   "Continental drift," *Sci. Am. 208,* No. 4 (April), pp. 86–100.
  1963b   "Hypothesis of earth's behaviour," *Nature 198*:925–929.

Wise, K. A. J.
  1964   "New records of Collembola and Acarina in Antarctica," *Pacific Insects 6*:522–523.

Wittmann, O.
  1934   "Die biogeographischen Beziehungen der Südkontinente," *Zoogeographica* (Jena) *2*:246–304.

Wodzicki, K. A.
  1950   *Introduced mammals of New Zealand* (Dept. Scientific and Industrial Research, Wellington, N. Z.).

Woollard, G. P.
  1962   "The land of the Antarctic," *Sci. Am. 207,* No. 3 (Antarctic number, Sept.), pp. 151–166.

Zeuner, F. E.
  1959   "Jurassic beetles from Grahamland, Antarctica," *Palaeontology* (Palaeontological Assoc., London) *1*:407–409.

Zimmerman, E. C.
  1948   *Insects of Hawaii, vol. 1, Introduction* (Univ. of Hawaii Press, Honolulu).

Index

This index is designed to be informative, not exhaustive. Minor references to names of places and organisms are not indexed. Names of persons are indexed only if the persons' work is discussed or if the names are not included in the Bibliography (pp. 219–229).